Experiments for
PHYSICS: MODELING NATURE

Important Notes

The contents of this book are adapted from our book *Favorite Experiments for Physics* and Physical Science. There are some uncorrected page number references, and some references in this book to pages or sections in *Favorite Experiments* that are not included in the present volume.

There are frequent references to Flinn Scientific as an equipment supplier. While Flinn Scientific is a fine supplier for schools, they do not serve home schoolers. For alternate materials sources, see the Materials List beginning on pg. 71.

Experiments for
PHYSICS: MODELING NATURE

John D. Mays

NOVARE
CLASSICAL ACADEMIC PRESS

Camp Hill, Pennsylvania
2022

NOVARE
CLASSICAL ACADEMIC PRESS

Experiments for Physics: Modeling Nature
© Classical Academic Press®, 2022
Edition 1.0

ISBN: 978-0-9966771-3-4

Classical Academic Press
515 S. 32nd Street
Camp Hill, PA 17011
www.ClassicalAcademicPress.com/Novare/

IS.01.22

Appreciations and Acknowledgements

As with my previous books, this book would not have been possible without the support, encouragement and collegial collaboration of the faculty at Regents School of Austin. Thanks especially to Chris Corley and Cathy Waldo, who continue faithfully to teach there.

Thanks to Caleb Kyle, who as a 15-year-old student introduced me to blowing up coffee creamer and taught me every single detail of performing that demonstration. Thanks to my old supervising teacher from the 1980s, Sam Saenz, who taught me the art of hunting monkeys.

Thanks to all my family for their continuous encouragement. Special thanks to Jeffrey and Rebekah for their direct support of and contributions to this project. And thanks to my brother-in-law Ray Arneson, who gave me the plywood Magic Belt hook.

Finally, thanks to good friend and gentleman scientist Dr. Chris Mack. To produce good science books one must never tire of infinitessimal improvements in the details. In this regard, Chris is just as much a perfectionist as I am, and never lets me off the hook.

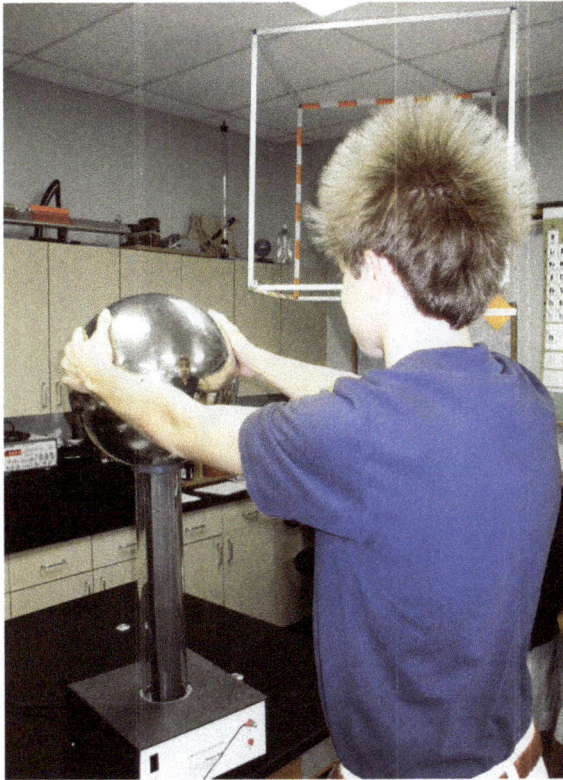

Contents

Why I Wrote This Book

Back in the 1980s when I began teaching ninth grade Physical Science and Senior Physics, I was at a public high school that had not invested much in apparatus for physics labs. So as I began planning my lessons for that year I did what anyone else would do. I got the "Lab Manuals" for our texts, studied them to learn what equipment was required for all the experiments, wrote up a big purchase order and bought all of it. You probably know what kind of equipment I am talking about—little carts; mass sets; ticker-tape timers; ramps and little wooden boxes for fooling with friction; spring scales; a balance beam for demonstrating torque; DC circuit kits; etc., etc.

I studied the teacher's guides, wrote up lesson plans, and enthusiastically led the students in their lab activities like I was Delacroix's *Liberty Leading the People*. My doubts about the whole process began when I tried to explain how to use the space between the dots on the timer tape to determine the speed of the cart at different times. What a cumbersome way to measure velocities, I thought. In a decade when everything was going digital and many students already had computers at home, the technology seemed crude and the unwieldy strips of paper were a pain. Moreover, it was often the case (and still is) that the apparatus either didn't work or the accuracy was low. The torque balance beam demo with a meter stick is so finicky it is frustrating to the teacher and unconvincing to the students. The wheels tend to come off of the little carts. The DC circuit kits look like they belong in an elementary school classroom.

The mathematical connection between the spacing of the dots and what we were studying was there, of course, but even after careful review with the students many of them were clearly thinking, "We do what? With what? Okay, whatever you say." Not a good way to build interest in what I felt should be everyone's favorite subject.

An effective classroom is not a place where students mindlessly go through the motions of tasks they do not understand in order to perform an "experiment" that does not interest them.

So I began using my imagination to dream up better ways to engage students in lab activities that would provide them with a more meaningful encounter with the basic principles of physics. My motivation was to develop experiments and demonstrations that would be academically solid, inexpensive and interesting.

The activities in this book are the results of those years trying things out and improving my home-made apparatus to increase the reliability and accuracy of the results. These experiments and teacher demonstrations are the ones I presently do in my own classes, the little carts and friction boxes now gathering dust in a closet.

Most of these experiments can be performed very inexpensively. In my descriptions I indicate how to do the experiment with little investment, making the experiments accessible to schools and homes with limited funds. Over the years I have enhanced some of these experiments with digital electronics for data collection. This makes the experiment more interesting to the students, who are surrounded with digital electronics and tend to find anything else uninteresting. The electronics also increase accuracy significantly, improving results and making the analysis more satisfying. But my experience has shown that the simple act of doing an experiment outside with a pickup truck is so exciting for the students that they will love it whether you collect force data with fancy digital equipment or with lowly bathroom scales purchased from a discount store, as I did for many years. If budgetary constraints are an issue for you, start doing the experiments without the fancy digital equipment. You can modify the experiment and add the electronics over time as funds become available.

I know there are a lot of books out there with ideas for science experiments. But the emphasis in this book is on experiments that are captivating, are low cost (at least initially), provide solid opportunities to do physics (and a little chemistry), and use equipment that is either already familiar or worth knowing about. I hope some of these experiments will enhance your own classes.

How Many Labs to Do

As a public school teacher in Texas in the 1980s I was required to make 40% of my class time laboratory-based. Even as a young man just beginning a teaching career I knew this was an insane standard. For starters, once students are in high school their science studies need to be academically rigorous. If a student finishes a course in physics, he or she ought to know the basic principles of physics and be able to solve problems. But three days per week to study the theory is not adequate time for students to master the basic calculations of high school physics.

Reinforcing this problem is the fact that typical "lab manuals" include 25 or 30 different experiments, implying that teachers need to spend a day or two every week running labs and that students need to be filling out report sheets every week. Such an environment, requiring students to race through lab activities and crank out weekly fill-in-the-blank reports will almost certainly be superficial, feeding what I call the *Cram-Pass-Forget cycle*. Students cram for tests, pass them, and then forget most of what the teacher wanted them to learn. I am opposed to superficiality in principle and deeply concerned about this deplorable trend in science education in particular. Students should learn fewer topics but learn them deeply through extended class discussion, in-depth problem assignments, and report writing that is engaging. The result will be deeper comprehension and retention of fundamentals. This in turn will allow students more easily to grasp and master more advanced topics in future studies.

These considerations suggest we should be looking at more like five to six full experiments with lab reports each year for high school science classes.

The Importance of Real Lab Reports

Writing lab reports is a crucial element of the science laboratory experience. To equip your students with the information they need to prepare high quality reports, I commend to you my resource for students, *The Student Lab Report Handbook*.[1] This volume contains all the information students need to use lab journals and prepare excellent lab reports, and can be considered a companion volume to this book. In the preface to *The Student Lab Report Handbook* I make the case for requiring students to produce their own typed lab reports

---

1   *The Student Lab Report Handbook*, by John D. Mays (2009). Available from Novare Science and Math at novarescienceandmath.com.

from scratch, rather than using pre-printed forms in a lab manual. Writing such a report is a significant undertaking and requires a lot more time than students can afford to give every week. Thus, it is important to select lab activities very carefully so that they support the goal of deep engagement. Experiments have to hit on key topics, they have to be spaced out so there is adequate time for report writing, and they have to have enough complexity to support deeper analysis. In physics this deeper analysis usually involves predicting results from theory, and comparing experimental results to predictions with properly designed graphs.

In the Student Instructions for most of the experiments in this book there are remarks addressing specific issues pertaining to the lab report for the experiment. These comments are based on the assumption that students are asked to prepare reports in accordance with the requirements presented in *The Student Lab Report Handbook*.

Do We Need This Much Detail?

After reading some of my experiment descriptions, you may be tempted to accuse me of being pedantic to a fault. So much detail! How can anyone remember all of these little details?! This guy is nuts! Well, this is what the real world of science is like—excruciating, fastidious, mind-numbing attention to detail, and to the elimination of every conceivable source of error. In every lab activity you perform, impress upon your students the paramount importance of attending to detail and employing painstaking care. In so doing you are teaching them part of the ethos of good experimental science.

Teacher Background

In this book the theoretical background for the experiments is quite abbreviated. Teachers who are new to teaching physics or who would like to read the theory behind the experiments in full detail should refer to a good text on the subject.

Accuracy, Precision, Significant Digits, Units of Measure, etc.

I assume many of my readers already know the difference between accuracy and precision, and how to deal with units of measure. However, it is also the case that teachers without a strong background in physics are sometimes recruited to teach it. Most of the experiments in this book involve predicting a physical quantity, measuring the quantity in the experiment, and comparing the predicted value to the experimental value. This process always involves measurements and computations, and these in turn always involve units of measure and everything else that goes with making measurements.

For those readers who may not be familiar with the details of these issues, who would appreciate some advice on what to cover in class, or who would like a short tutorial on the units involved in the experiments in this book, I have included an Appendix that contains a primer on measurement.

Learning Objectives for a Secondary Science Laboratory Program

There are many learning objectives to consider when organizing a lab program for middle and high school students. Most of these objectives are realized over a period of several years, as students go through several different science courses and engage in a number of experiments in each course.

The general objectives I have identified and seek to address in the experiments in this book are listed in the table below. The goal is that after having completed the secondary course of study at a school, students will be competent in each of the objectives listed. The objectives listed in the table are addressed by nearly every experiment in this book.

General Learning Objectives for a Secondary Science Laboratory Program	
After completing the program of laboratory exercises in the secondary program, students will be able to demonstrate competence in each of the following tasks:	
1	State and follow standard laboratory safety practices.
2	Correctly identify and use standard laboratory apparatus.
3	Use proper care in setting up apparatus and handling materials to maintain a safe environment, protect equipment and maximize accuracy in results.
4	Describe and follow the proper methods for making measurements with common instruments. This includes identifying the types of errors that can introduce inaccuracies in measurements and describing how to avoid them.
5	State the role of precision in taking measurements, and relate this to the significant digits in a measurement.
6	Apply the scientific method to conducting experiments and to writing reports.
7	Apply appropriate logic to conducting experiments and to writing reports.
8	Maintain a proper lab journal.
9	Clearly explain the theoretical background behind an experiment using quantitative analysis where appropriate.
10	Use quantitative predictions from scientific theory to form testable hypotheses.
11	Clearly and efficiently describe a scientific procedure and the results and discoveries that followed.
12	Use appropriate care in experimental procedures and data collection.
13	Present calculations and data in a clear, organized fashion such that others can verify calculations or check results. This includes development of tables and graphs using standard scientific units and formatting.
14	Apply quantitative analysis to experimental data as appropriate.
15	Apply qualitative analysis to experimental results as appropriate.
16	Estimate uncertainty in measurements.
17	Apply cogent reasoning to analysis and discussion of experimental results. This includes reasonable identification of the factors that contributed to the difference between predicted and measured results (aka, "experimental error").
18	Use computer tools to take data, graph data, manipulate data and develop reports.
19	Use clear, concise, and accurate language in a technical style in scientific reports.
20	Explore the uses and limitations of unfamiliar scientific equipment.
21	Cooperate with team members successfully to accomplish each of the above objectives.

In addition to these general objectives, each experiment has one or more unique features that suggest specific objectives that apply to that experiment. These specific objectives are listed at the beginning of each experiment.

Student Instructions for Experiments

Student instructions are included at the end of each of the 11 experiments. These instructions may be reproduced and distributed to students. Alternatively, PDF files of the student instructions are available as free downloads from our website, novarescienceandmath.com. These may be downloaded, reproduced and distributed to students. Simply go to the Free Resources tab on the website and enter the pass code "novarefavexp."

A Note About Experimental Error

One of the conventional calculations in high school science labs is the so-called "experimental error." This experimental error is typically defined as the difference between the predicted value and the experimental value, expressed as a percentage of the predicted value, or

$$\text{experimental error} = \frac{|\text{predicted value} - \text{experimental value}|}{\text{predicted value}} \times 100\%$$

From the perspective of the average high school student, this use of "experimental error" makes perfect sense. After all, student are studying well-established theories and the goal of the experiment is to learn about the theory, not to validate or refute it. In the world of science, however, experiments are the golden standard by which theories are judged. When there is a mismatch between theory and experiment, it is often the theory that is found wanting. That is how science advances.

In my early books, such as *The Student Lab Report Handbook*, I used this same terminology ("experimental error") to express the difference between prediction and result. Over the years, however, research and discussions with practicing scientists have led me to the conclusion that this terminology is misguided. Used in this way the term *error* implies that the theory is *correct* and that the error in the experiment may be summarized by this difference equation. However, the difference between prediction and experimental result may not be caused by deficiencies in the experiment. In more general scientific practice the theory may *not* be correct. Thus, in secondary classrooms it is better to reserve the term *error* for discussions about lack of accuracy in specific measurements, when the measurement is known to contain or is suspected of containing error (that is, differing from the true value, see Appendix). Referring to the overall difference between prediction and experimental result as "experimental error" is a bad habit to get into.

Consider this case: an experimental measurement of velocity produces a value that is consistently less than the predicted value. Most likely this is because the predictions did not take air resistance into account. Is this an experimental error? It is more correct to say that the theory is inaccurate because we made the unrealistic assumption that there would be no air resistance. Such causes of differences between predictions and measurements are quite common, and it is great if the future scientists in your class can understand that this is not an error in the experiment.

As a result of these considerations, beginning with this book I am adopting a different convention. Henceforth I will use the phrase "percent difference" to describe the value computed by the above equation. When quantitative results can be compared to quantitative predictions, students should compute the percent difference as

$$\text{percent difference} = \frac{|\text{predicted value} - \text{experimental value}|}{\text{predicted value}} \times 100\%$$

In the Discussion section of their lab reports, students should state the value(s) of the percent difference for their experiment. After doing so, much of their subsequent analysis of the experimental result will consist in attempting to identify the reasons for this difference. Students may use the possibility of *errors* in different measurements, along with other factors such as lurking variables or insufficiently elegant experimental methods, in their attempts to account for the prediction-result difference. Being able to explain the percent difference for an experiment is one of the most important jobs of the scientist—and the science student.

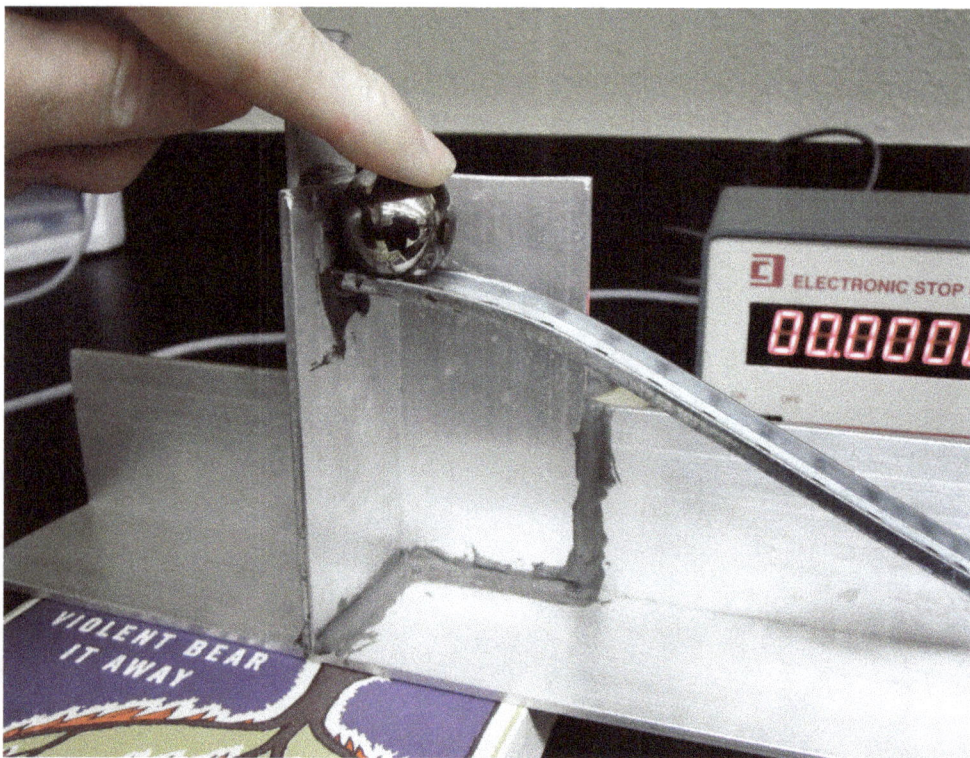

The five experiments in this section were developed for upper-level high school physics students (typically seniors, but sometimes juniors). If you teach senior physics you may have turned straight to this section. I encourage you to read the Introduction for the background on why I developed these experiments and what I was trying to accomplish by departing from the standard sorts of experiments high school students usually perform.

I also encourage you to read through the six experiments in Part 1 of this book, which I use with ninth grade students. When I teach senior physics I assume these six experiments were all performed by the students in their freshman year, thus providing background experience and context for our experiments in senior physics. For example, Experiment 2 (Newton's Second Law) is an excellent investigation which I have done with older students. Over the years I decided that it was accessible for younger students, so I moved it to the ninth grade in our curriculum. But if your students have not performed it in a prior science class I recommend you include it in the program of experiments you do with your senior class. The same thing goes for Experiments 3 (Conservation of Energy) and 4 (DC Circuits). These experiments treat standard topics in physics all students should experience.

As I wrote in the Introduction, the ability to write a solid lab report from scratch is one of the important tools students use to engage in science studies. While writing the material for the experiments in this section my operating assumption is that students engaging in these experiments will be writing full reports, and I have written the student instructions with this in mind.

Finally, please see the Introduction (page 6-7) or the Appendix (page 249) for information about my use of the phrase, "prediction-result difference ratio."

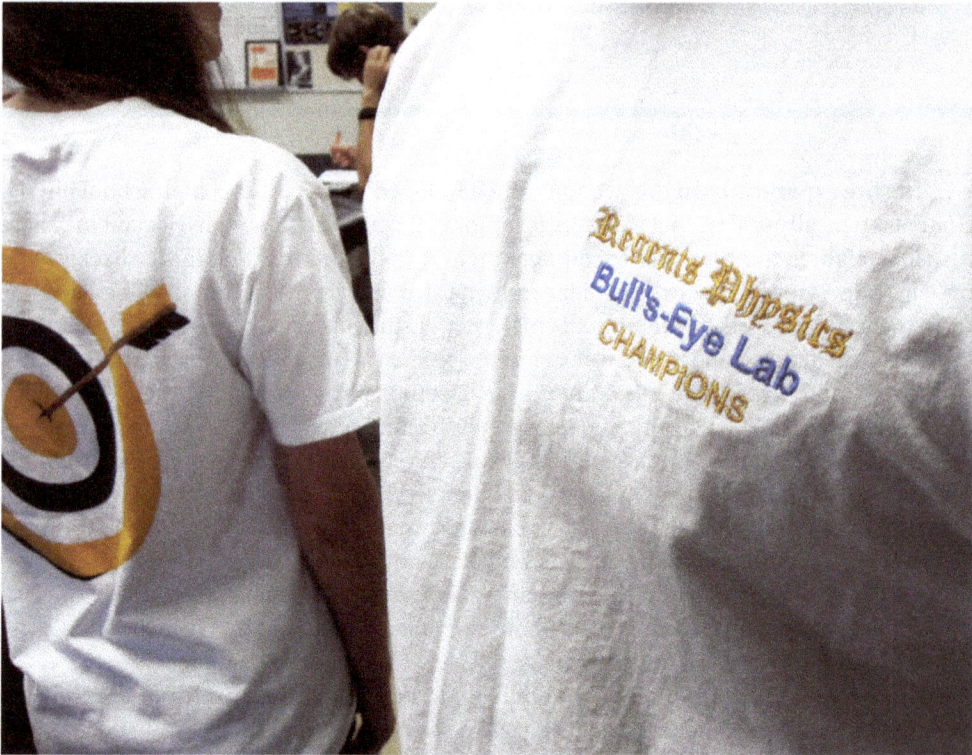

Experiment 1   The Bull's-Eye Lab

Learning Objectives

Features in this experiment support the following learning objectives:
1. General objectives for laboratory experiments (see page 4).
2. Use vector-based equations for two-dimensional projectile motion to make predictions.

This is our first experiment of the year in physics. The experiment is both simple and fascinating for the students. I make it doubly fun by having the teams compete with one another for a team prize. The team that gets the lowest prediction-result difference ratio wins a Bull's-Eye Lab Champions T-shirt for each team member. Moreover, I obtained permission from the school administration for the champions to wear their T-shirts to school on Physics exam days for the rest of the school year. (This is viewed as an especially valuable prize since students at our school are ordinarily required to wear school uniforms every day!) Thus, Bull's-Eye Lab day each September is full of anticipation, tension, and the glory of victory.

Materials Required (per team)

1. steel ball, 1 inch diameter, such as No. AP5626 available from Flinn Scientific (flinnsci.com)
2. laboratory support rod or ring stand
3. clamp (small)
4. shelving support rail, 5/8 in wide x 3/8 inch deep (available at hardware stores)
5. stop watch
6. masking tape
7. plumb bob (available from hardware store or construction supply store)
8. nylon string
9. meter stick
10. carpenter's level
11. target (photocopied)
12. carbon paper (one sheet for the class)

Experimental Purpose

Use the principles of projectile motion to predict where a rolling steel ball will land when it rolls off a table and hits the floor, and compare the prediction to the actual landing spot.

Overview

Students assemble a makeshift ramp/track on a horizontal table. Then they release a 1-inch diameter steel ball so it will roll down the ramp and off the table. Between the ramp and the edge of the table students mark off a timing zone 80 cm or so long. By timing the ball several times while it is rolling on the table through the timing zone they determine the velocity the ball has when it leaves the table. Using this velocity and the height of the table, they predict where the ball will land when it hits the floor. They work out this prediction during the lab time, and then tape down a target on the floor with the center at the predicted landing location. When they are ready for the moment of truth, the instructor brings a sheet of carbon paper and places it carbon-side down on the target. Then the team releases the ball. When the ball hits the carbon paper it leaves a distinct black dot on the target where it landed.

This experiment can be performed very inexpensively. For the ramps you can purchase one or two sections of the support rail used in adjustable-shelf bookcases and cabinets. These sections of rail come in six-foot lengths. Cut the rail so that each team has a piece about 22 to 24 inches long. Use a grinder to grind off one end of the rail at a steep angle so the upper edge of the track where the ball will be rolling comes down close to the table top when the track is angled at about 10 degrees (see photos). After completing the rough grinding, be sure to smooth off all sharp points and edges with a file.

Pre-Lab Discussion

Perform this experiment after students have spent several days solving projectile motion problems. You will not then need to spend any time explaining how to calculate where to place the target. On the day of the experiment I briefly review how to obtain the initial (horizontal) velocity of the ball by timing it as it rolls through the timing zone on the table. I advise students to time it several times, with different students operating the stop watch, and to use the mean time to estimate the ball's initial velocity.

I generally assume my students are honorable, but I also wish to avoid any team gaining an unfair advantage by "accidentally" allowing their ball to hit the floor during the time trials, perhaps noticing approximately where it landed. Thus, I have an important rule for this experiment that I enforce with uncompromising rigor: Any team that allows their ball to hit the floor for any reason prior to the official run for the target forfeits 25 points from each of the team members' lab reports. This rule has been effective in motivating each team to keep their ball on the table until they are ready for their official run for the target.

The rest of the pre-lab discussion needs only to focus on a few practical details.

1. Caution the students to make the angle of the ramp fairly low, as indicated in the accompanying illustrations. This is the only way to keep the speed of the ball low enough so that it can be accurately timed in the timing zone with a stop watch.

2. There needs to be a convenient way to consistently position the ball on the ramp. The little clamp used to attach the ramp to the support stand (see photos) will meet

    this need if it is positioned so that the metal arm of the clamp can act as a stop for the ball at the top of the ramp.

3.   Students need to use a carpenter's level to check the table to assure that it is level in both horizontal directions. This is critical, since even a slight tilt to the surface of the table can have a significant effect on the velocity of the ball.

4.   The bottom of the ramp should be secured with masking tape so it doesn't shift around.

5.   Students need to project the edge of the table down to the floor for measuring out to where the target needs to be placed. A plumb bob on a string is the easiest way to accomplish this.

6.   Students may need to be reminded that the variable they are predicting in this experiment is the horizontal displacement of the ball past the edge of the table. Accordingly, when calculating the prediction-result difference ratio the distance from the landing spot to the center of the target, which is the measurement everyone is immediately interested in, is not the value to use. Instead, the actual total horizontal travel is compared to the predicted value. The horizontal displacement depends on the initial velocity of the ball and the height of the table, and in my lab the predictions are typically in the range of 30-35 cm. Thus, missing the target by 1.5 cm long or short represents a prediction-result difference ratio of about 5%.

7.   For reasons explained below, I ignore any deflection of the ball to the right or left of the target and consider only the difference ratio in the direction of motion. Even though my reasons for this approach are due to the design of my lab tables, I think I would do it even if my tables were different. This is because the prediction the students are making is in the direction of motion, so that is where the difference ratio calculation is relevant. Of course, in the real world the details of mechanical design also play a major role in how well engineered systems perform, so you would be justified in taking left-right deflection into consideration in determining the winner of the prize.

Additional Experimental Details

1.   The ramps are made of the metal railing used inside of bookcases to support the shelves. This material is very inexpensive and makes a perfect track for a 1-inch steel ball. As mentioned above, angle the ends of the rails for a smoother transition for the ball as it moves from the rail to the table top. The ball will still bounce a bit when it hits the table, but not enough to cause a problem. Make sure to go over all of the edges of the metal with a file to remove burs and sharp corners.

2.   As you can see in the illustrations, the tables in my lab have sockets into which support rods may be inserted. I like this feature because the support rods can't get knocked over like a ring stand can, and storing the rods is easier than storing ring stands. However, this design does present a limitation for this experiment: There is a hole for another support socket right in the runway for the rolling ball! We get around this by placing the ramp at a very slight angle so the ball misses the hole.

This results in a small deflection to the left or right, but the effect on the horizontal displacement in the direction of travel (the main variable) is negligible.

3. The small angular deflection in the path of the rolling ball necessitated by the extra socket holes in my lab tables is one of the reasons why I have allowed students to base the prediction-result difference ratio only on the distance in the intended direction of travel (call it x) from the landing spot to the center line of the target. The other reason is that the experimental variable the students calculate is the *horizontal displacement* (x displacement) while the ball is airborne. So it is only the difference between this predicted horizontal displacement and the actual horizontal displacement that has meaning in the difference ratio calculation. (The predicted value of the y displacement, that is, the displacement perpendicular to the intended direction of travel, is zero.) Of course, displacement in the y direction will be captured by increased error in the x direction, which is incorporated into the prediction-result difference ratio calculation.

4. Team members should all agree on the calculations, and the horizontal displacement prediction, before their official trial is performed. Any discrepancies should be resolved in advance. This will help avoid a large prediction-result difference ratio due to error in the calculations. If a team's difference ratio is larger than just a few percent there is a reason for it, and team members should strive to identify it and discuss it in their lab reports.

5. Since carbon paper is a vanishing commodity, you may have trouble finding it at the local office supply store. There are still plenty of places that have it online, but you may have to buy a packet of 100 sheets or so. (One or two sheets is probably enough to last your entire teaching career!)

Student Instructions

A set of instructions you may reproduce and give to students begins after the following illustrations.

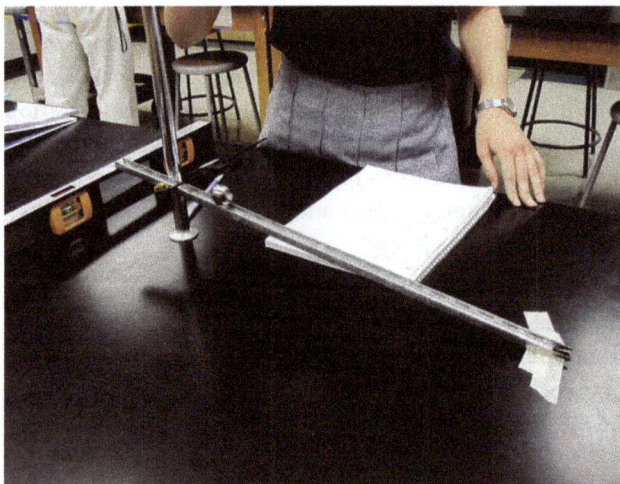

The basic setup, with the steel ball on its way for a time trial.

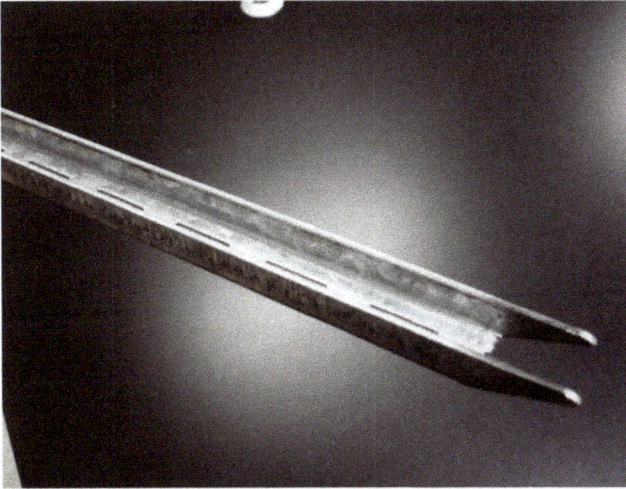

The end of the rail is ground at an angle to help smooth the ball's transition from ramp to table top.

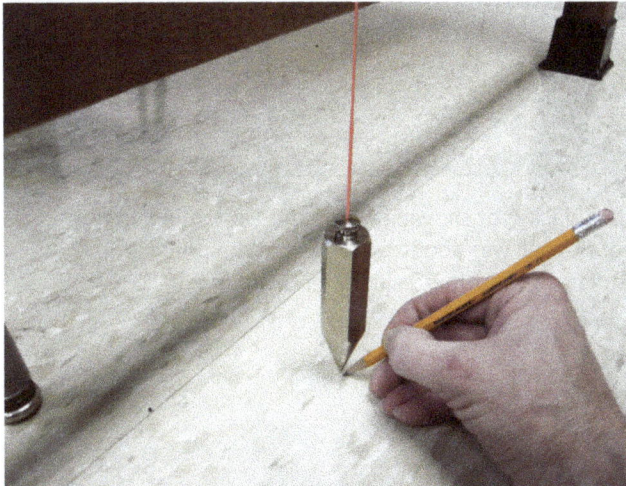

A plum bob hanging off the end of the table is used as an aid in projecting the edge of the table down to the floor. The ball's horizontal displacement while in flight is measured from this mark.

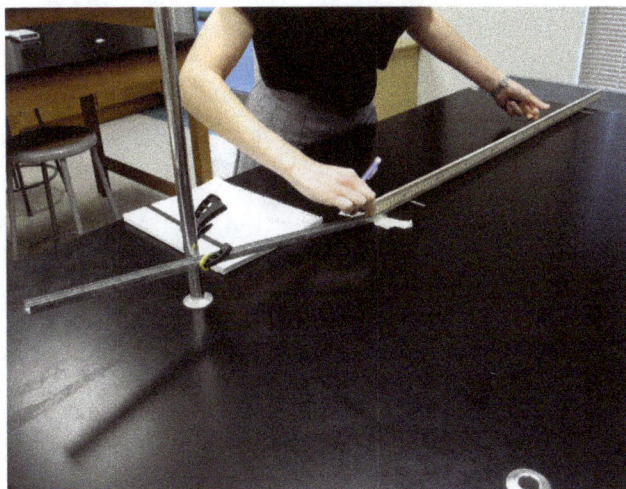

Laying out the timing zone.

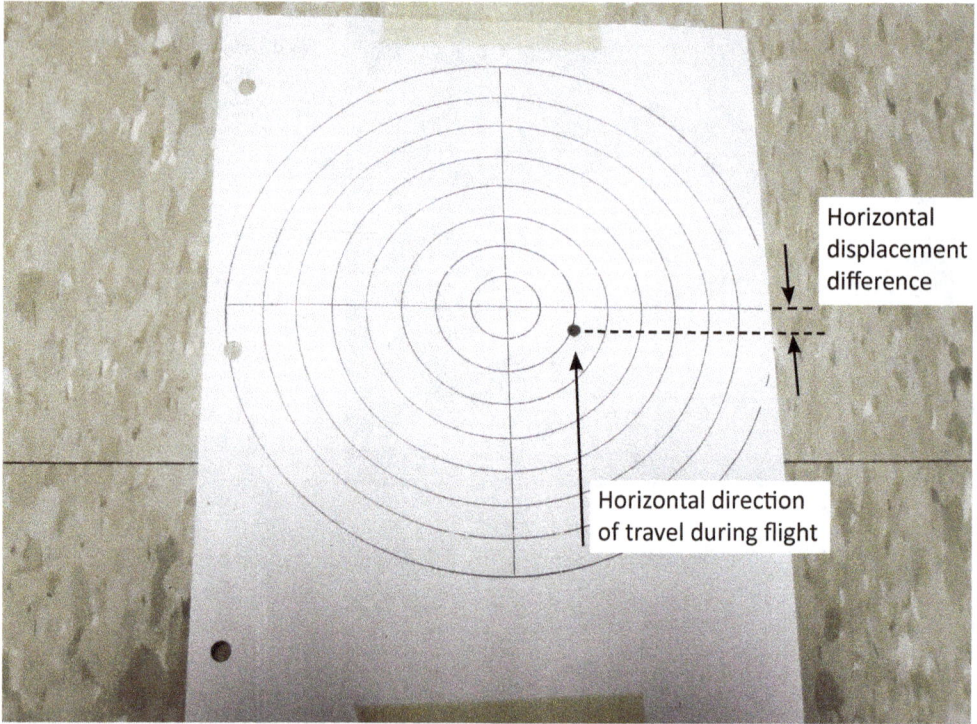

The target taped in position, with the black spot showing where the ball landed. In this case, the horizontal displacement of the ball was about 1 cm short of the prediction. For a prediction of 30 cm, this would result in a prediction-result difference ratio of 3.3%.

Front and rear views of the coveted prize for the winners—Bull's-Eye Lab Champions T-shirts!

The Bull's-Eye Lab
Projectile Motion

Your task is to make an accurate prediction of where a steel ball will land when it rolls off a table top. You will do this by setting up a ramp for the ball to roll down, and then timing the ball as it passes through a marked timing zone on the table top. You will use the time data to determine the horizontal velocity of the ball on the table top. This velocity, combined with the height of the table top, will enable you to calculate where the ball will land. After you have calculated and marked where the ball will land, you will tape a target onto the floor with its center placed precisely at the predicted landing spot. When you are ready, your instructor will place a sheet of carbon paper face down on your target. You will then let your ball roll down the ramp and off the table, and the carbon paper will mark where it lands. Your grade will be based on (a) how close you got to the bull's eye, and (b) your report, including your analysis of your results and errors.

SPECIAL WARNING

You will only get one chance to hit the bull's eye. Do not let your ball roll off the table onto the floor until you are ready for your "official trial" which must be witnessed by your instructor. If any group lets their ball roll off the table and onto the floor, even if by accident, the members of that group will each incur a penalty of 25 points on their lab reports.

Experimental Procedure

1. Using the stand, ramp and clamp, set up a ramp on your table. Roll your steel ball down the ramp several times to get an idea of what its velocity will be as it leaves the table. A small ramp angle will allow the ball to move slowly enough so that timing its motion on the table top may be done with reasonable accuracy.

2. Mark off a timing zone between the end of your ramp and the edge of the table. In this zone place two strips of masking tape. These strips will be used to time the ball as it goes through the timing zone so that you can calculate the velocity with which it will leave the table using $v = d/t$.

3. Use the ramp clamp as a stop at the top of the ramp for starting the ball consistently.

4. *Without letting the ball hit the floor*, use a stop watch to time the ball several times through the timing zone, after it has been released from the top of the ramp. Average these times and use this average along with the length of the timing zone to calculate the ball's velocity as it leaves the table. Be sure to record all data in your lab journal.

5. Measure the height of the table and use this with your calculated initial velocity to calculate where the ball will land when it hits the floor.

6. Use a plumb bob to project the edge of the table down to the floor. Make a small mark with a pencil on the floor locating the edge of the table. From this mark, measure the predicted horizontal displacement for the ball's flight. Mark the spot on the floor where the ball will land. Tape your target down, centered on that spot.

7. You need to devise a way to assure that the center lines of the target are parallel/perpendicular to the ball's direction of travel. If the floor surface is commercial tile, the lines in the tile can be used to square up the target. Otherwise, you may need to project the table edge down to the floor in two places and measure out from each of them to locate the center line of the target.

8. Notify the instructor that you are ready for your official trial. The instructor will place a piece of carbon paper face down on the target. Then your team will let the ball roll freely down the ramp and hit the target.

Notes on calculating the prediction-result difference ratio for this experiment:

1. As always when comparing theoretical predictions to experimental results, you must calculate your prediction-result difference ratio. Your predicted value is the horizontal distance the ball will travel while it is in the air. Thus, your experimental value is the actual horizontal distance it traveled while in the air. This is a different number from the distance your ball was away from the center line of the target.

2. Your horizontal distance is the distance to the target or the landing spot from the line of the edge of the table projected onto the floor. Do not try to account for angular error in the trajectory, which might have made the ball land to one side or the other of the target. Discuss only the straight-line distance error to the line of the edge of the table.

Reproducible Target

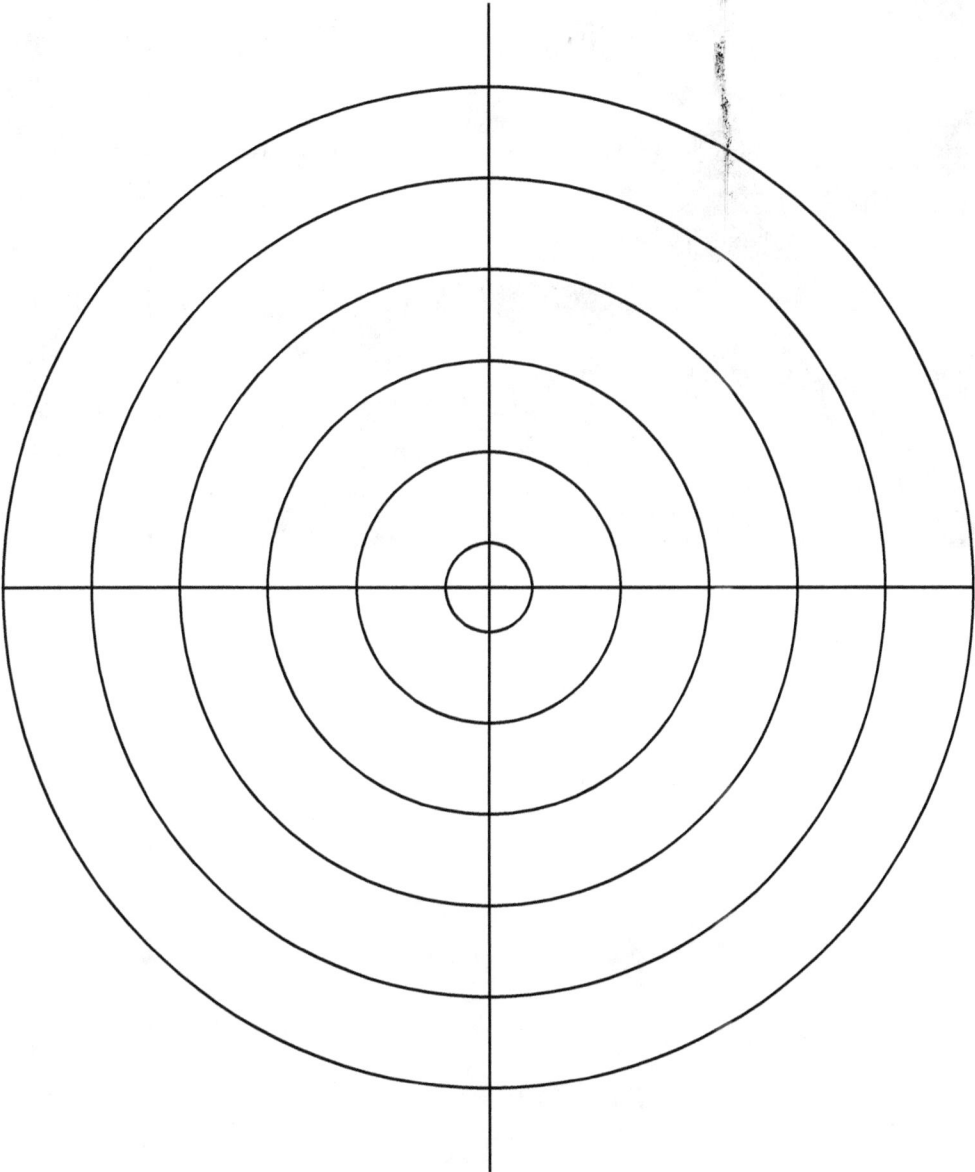

Learning Objectives

Features in this experiment support the following learning objectives:
1. General objectives for laboratory experiments (see page 4).
2. Develop a reliable, original experimental approach that will allow accurate and precise determination of four separate variables.
3. Explore and learn to use unfamiliar scientific equipment.
4. Use correct polishing techniques to prepare metal surfaces.
5. Work within a project budget.
6. Accomplish interim experimental tasks and present results in interim reports according to a project schedule.
7. Work with an experimental team to brainstorm, plan and execute a multi-phase experimental project.
8. Use computer tools to calculate the standard deviation as an estimate of uncertainty.

This project requires student teams to develop their own experimental design to measure the static and kinetic friction coefficients, μ_s and μ_k, for metal-on-metal contact, with and without lubrication (four separate coefficient values). Students develop and execute their experimental plan just as they would if they were involved in an engineering project in industry, with deadlines, a budget, access to a finite variety of laboratory resources, and freedom to use their ingenuity and creativity to solve the problem any way they can. If students can procure equipment and materials at no cost, then they don't count against the budget. A set of the two brass parts (a section of plate and a piece of flat bar) is furnished to each team, along with a supply of metal polish, and the cost for all of these items is charged against each team's budget. Other stock laboratory materials may made available to teams, depending on what equipment the lab has on hand.

The beauty of this project is that students are given no advice on how to go about determining the values of the coefficients. So their thinking will begin with the basic definitions of the friction coefficients. But soon they will be thinking in terms of vector force analysis and kinematics, and asking themselves a chain of questions about how measurement of one thing can lead to the determination of another thing, and so on, until they finally have a way of getting at the coefficient itself. When this happens, this simple experiment about friction will transform into a nice workout in the computations associated with kinematics and dynamics.

Materials Required (per team)

1. brass plate, 10 in x 4 in x 5/16 in (approx.), may be sourced from Industrial Metal Supply, (industrialmetalsupply.com)

2. brass flat bar, 1 in x 1 in x 1.25 in (approx.), may be sourced from Industrial Metal Supply, (industrialmetalsupply.com)

3. waterproof polishing paper in four grades: 120-C, 220-A, 320-A, and 400-A. These are available from Abrasive Sales.com (abrasivesales.com) as part nos. 19823, 19808, 19798, and 19795, respectively.

4. cleaning cloths

5. nylon cord

6. WD-40 spray silicon lubricant

7. laboratory balance

8. other standard laboratory materials as available in the laboratory, such as low-friction pulleys, table clamps, mass sets, adjustable ramps with angle gauge, timing equipment, and measurement tools.

9. other materials furnished by team members

Experimental Purpose

Design methods to produce precise, accurate measurements of static and kinetic coefficients of friction (μ_s and μ_k), and implement these methods to measure the values of μ_s and μ_k for brass-on-brass contact under dry and lubricated conditions.

Overview

I developed this experimental project around using two brass parts so that students would investigate coefficients associated with metal-on-metal contact. Clearly, the same idea could be applied using two pieces of wood or other materials. In fact, using wood or other materials would cost a lot less and entail less hassle. However, it seems to me that the use of metals is really crucial for making this investigation a success for the students. Simply put, using common pieces of wood seems immediately to be *boring*, while working with brass parts seems *intriguing*. Thinking about this difference has persuaded me that there is an opportunity here to enhance student interest and provide them with an experience that will prove especially valuable if they enter careers in science or engineering. Using metals brings in such considerations as these:

1. Many students have never worked with solid brass materials before. There is added interest simply in the novelty of working with unfamiliar materials. When presented with the brass pieces the students' first impulse is to pick them up and handle them, feeling the density, etc.

2. By handling these materials they will learn first hand about the density of this alloy and its susceptibility to scratching due to its softness relative to steel.

3. As they polish the parts to remove scratches and oxides, students will be fascinated by the colors and the change in appearance of the brass.

4. Wood is seldom used in mechanical or machine design, whereas metals are universally used. Thus, measuring friction coefficients for metal-on-metal contact is

a more realistic scenario for application of the physics involved, and students are subconsciously aware of this.

5. The coefficient of static friction for dry metal-on-metal contact can be quite high, and, counterintuitively, the more polished the materials the higher it gets due to the Coulomb attraction between planes of atoms. This phenomenon should generate a good deal of classroom discussion about the nature of friction, its cause, and the consequences for mechanical design.

Students will acquire four sets of data in this project, static and kinetic friction coefficients of brass-on-brass, both with and without lubricant. Each data set will consist of at least six separate measurements, providing an opportunity for teams to determine and report the uncertainty (s) values in their data. Low values of uncertainty will contribute favorably to the students' "performance reviews," that is, the grades on their reports. High precision in the measurement techniques will allow variation in the measurements to be measurable, while high accuracy will reduce uncertainty. The description of the experimental method in the report and comparisons between the teams' reports should allow the instructor to judge how accurate the results are.

From the perspective of educational value, three major elements are present in this project. First, the students must develop their own experimental design. This immediately takes the experiment out of the realm of the canned physics experiment and places it in the more exciting realm of industrial research. In today's technology-driven culture just about anyone can imagine participating on a design team of this nature.

Second, students will need a thorough understanding of how friction works in a mechanical system (at the macro level, as opposed to the microscopic level), including the coefficients and equations for both static and kinetic (sliding) friction. They will also need to be comfortable with the basic calculations of kinematics and dynamics, including vector force analysis for blocks sliding on inclined planes with friction, Newton's Second Law, and so on.

Third, students will use a simple spreadsheet to facilitate data analysis. Familiarity with spreadsheets is a significant advantage for students entering science or engineering programs in college.

Measuring the coefficient of static friction, μ_s, is not difficult. The team merely needs to devise a way of applying a measurable force to the brass block parallel to the surface of the plate (with the plate fixed horizontally). Then the amount of this force must be incrementally increased until the block breaks free and moves. The weight of the block and the force required to make it move can be used immediately to calculate the value of μ_s.

Measuring the coefficient of kinetic friction, μ_k, is much trickier. Essentially, the measurement requires students to find a way to make the block move at a constant speed, and to measure the force necessary to do this, which would be equal in magnitude to the force due to friction. If the setup depends on the block moving at a constant speed, then students must devise a means of verifying that the block is in fact moving at a constant speed, which is not a simple task. Alternatively, and perhaps more easily, students can devise a setup that results in a uniform, measurable acceleration. Once the acceleration has been determined the friction force may be calculated using standard dynamics calculations.

If electronic timing equipment is available (infrared sensors and digital timers, such as described in the Hot Wheels Lab in Part 1, Experiment 3) additional possibilities open up; if not, the teams will need to do without it. It may or may not occur to students that a movie taken with an iPhone can be imported into iMovie and analyzed to extract time data. Since many students now have access to these tools, there is no real obligation for the science lab to purchase expensive timing equipment just to make this project tractable.

The availability of electronic force measurement equipment would seem to make the kinetic problem easier, but even then students will have to determine a way of creating a constant force that will pull a system at either a verifiably constant speed or with a known acceleration. If they decide to judge by eye that the system is moving at a constant speed, that is up to them, but the crude method of pulling an object by hand at a constant speed with a spring scale should be disallowed as not being nearly accurate or precise enough.

The process of thinking though the possibilities of an experimental design will have students buzzing about not only the definition of the coefficient of friction, but also about various combinations of ramps, pulleys, masses and timing apparatus. Teams will probably need to try out several different ideas for measuring the coefficients until they arrive at a solution. Because of this, it will be good if each team has a dedicated work space available at specific times where they can test ideas. If teams are held accountable for producing their own original ideas, privacy and security will also be a concern.

Depending on the school's resources, equipment available to students for this project could be extensive or sparse. If the inventory is sparse students will have to locate their own apparatus, which can make the project both more challenging and more interesting. If the lab is equipped with a supply of standard apparatus, these items can be made available to students. If the lab has the funds for such equipment, I recommend making the following items available to each team:

1. Inclined plane with an angle gauge capable of indicating the angle to at least three significant digits.

2. Low-friction pulleys with mounting clamps that will fit the raised end of the inclined plane.

3. Mounting clamps for attaching the low-friction pulleys to the edge of a table top.

4. Mass set.

Project Schedule

Experienced teachers know that when students are given a project assignment that stretches out over several weeks they are strongly tempted to procrastinate. Students are quite often driven by the closest deadline and tend to ignore everything else. In a project of this kind students do not really know how much time the project will take, and are highly likely to underestimate the time required for trial and error. Thus, procrastination will almost certainly result in poor work hastily performed at the last minute, or even failure to complete the project. To avoid such disasters, give your students a set of interim project milestones with due dates. This will force them to keep moving. I also recommend that you allocate several class days during the term for teamwork on the projects. Much of their work will need to be performed outside of regular class time, but a few class days devoted to working on this project will really help (as long as they are not allowed to use that time studying for tests, etc., which they are likely to attempt).

The table below shows a sample schedule I used in the fall of 2011. The schedule is based on presentation of friction in class in late September, with project assignments distributed a few days later and final lab reports due from each student prior to end-of-term exam preparations.

Project Milestones and Critical Dates	
Date	Activity and/or Submittal Due
Wednesday, Oct. 5, 2011	First in-class lab session. Teams begin working on experimental plan and identifying needed materials.
Friday, Oct. 21, 2011	Second in-class lab session. Teams continue planning and begin experimenting with possibilities for actual setup.
Friday, Oct. 28, 2011	Experimental plan due (typed).
Thursday, Nov. 3, 2011	Third in-class lab session. Teams complete setting up and begin taking data.
Friday, Nov. 18, 2011	Data summary due (typed).
Tuesday, Dec. 6, 2011	Final reports due.

Notes on Metal Materials

The small brass pieces required for this project will be cut from larger stock by the metal supplier. The term *plate* refers to metals originally fabricated in large sheets. *Flat bar* is the term used to described materials manufactured in long bars around 20 feet in length. The 10 in x 4 in x 0.25 in piece may be cut from either plate or flat bar, whichever is most readily available and thus least expensive. The 1 in x 1 in x 1.25 in block will be most economical if you specify a piece of 1 in x 1 in flat bar 1.25 inches in length. When I first ordered these parts I ordered several sets so that I could have several separate student teams working simultaneously. (In a class of four students or fewer I would have them work together as a single team.) When the parts arrived I took them down to a local machine shop and had 1/8-inch holes drilled through the centers of the blocks in two directions (see photo on page 178). This

Wet-polishing the brass.

provides students with a convenient way of attaching a cord to the block for testing. During the testing the block is oriented so that one of the 1 in x 1.25 in surfaces without a hole is in contact with the plate underneath.

The brass metal parts specified in the materials list will come from the metal supplier with sharp edges. Go over all the edges and corners with a flat file so that there are no sharp edges on which students can potentially cut themselves. Be careful not to touch the steel file to the working surface of the brass, because the lower hardness of the brass means it will be easily scratched or gouged by contact with steel.

As they come from the supplier, the brass parts will be covered with many surfaces scratches, easily seen from the photo on page 178 and in the upper section of the plate shown in the photo on the next page. All these scratches add up to a rough surface that will affect the experimental results. The first year you perform this experiment you can furnish each team with the materials to polish the parts to the point where the individual scratches are not visible to the naked eye. The polishing of one side of the plate and one surface of the block will require an hour or so of the students' time the first year the parts are used.

As illustrated in the photo on the previous page, place the plate under a gentle stream of running water. Using firm pressure and a swirling motion, polish the plate with four separate grades of waterproof polishing paper, beginning with 120 grade, and moving through 220, 320, and 400 successively. When you see the 120-grade paper you will probably think that it looks way too rough for the job, but to remove the

The polished area contrasts sharply with the unpolished area.

deeper scratches and pits this grade will be necessary. Always use an orbiting, swirling motion while polishing; never polish by moving linearly back and forth. Polish the entire working surface of the plate and the block. The polishing process will require about five minutes for each grade of polishing paper for the 4 inch x 10 inch plate.

After all four stages with the wet polishing paper have been completed, give the entire part a wash with dish soap and water.

In the photo to the left, the upper section of the brass plate is unpolished, just as the plate came from the metal supplier. In the center and lower sections the plate has been polished with the sequence of different grades of paper, ending with the 400-grade paper.

When storing the brass parts, spray each part down completely with WD-40 to prevent oxidation. Wrap the parts in cleaning cloths to protect them from scratches. The following year when you are ready to do the experiment again, have the students wash the parts with soap and water to completely remove the WD-40. After drying, have the students go over the working surfaces with the 400-grade polishing paper to renew the surfaces. If there are noticeable scratches or gouges in any working surfaces students will need to repeat the entire polishing procedure.

Pre-Lab Notes

This experimental project will take place over an extended period of time, rather than in a single class period, so there is no particular pre-lab discussion specific to the project except for standard instruction on friction, normal forces, force vectors, inclined planes and the like.

Students should be made aware of the following issues:

1. Students should polish the metal surfaces that will be in contact during testing to remove surface scratches visible to the naked eye. Carefully clean with soap and water after polishing.

2. For the non-lubricated trials, make sure the brass parts have been carefully polished and cleaned with soap and water to remove any residue of lubricant.

3. For the lubricated trials we are using a light coating of WD-40.

4. Students must be very aware of sources of friction in their apparatus other than the friction under investigation (such as in pulleys), and take active measures to minimize such friction as much as possible.

References

This experiment is a great opportunity for instructor and students alike to become more informed about the technical terms and processes associated with metal working. Check out the articles at wikipedia.org entitled "Polishing (metalworking)," "Friction," and "Hardness."

Student Instructions

A set of instructions you may reproduce and give to students begins after the next photo.

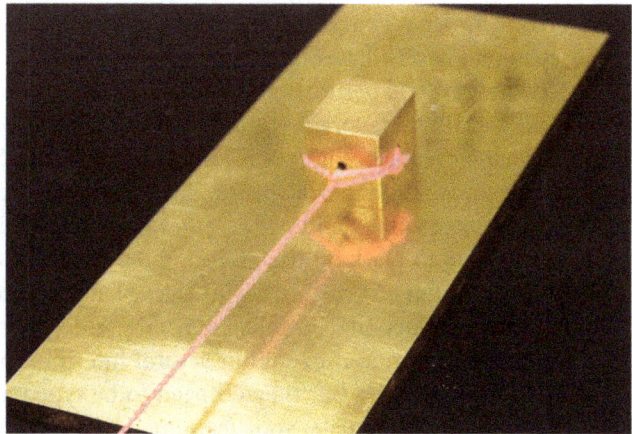

The nylon cord tied through the brass block and positioned for pulling.

The Friction Challenge
Coefficients of Friction and Experimental Design

For this project you will work in a team to develop your own experimental design. Your goal is to devise methods for measuring the static and kinetic coefficients of friction, μ_s and μ_k, for brass-on-brass contact, with and without the presence of a lubricant. You must measure each of these four coefficients in at least six separate trials, and use the sample standard deviation (s) of each data set as an estimate of the uncertainty in your experimentally determined values for μ. The experimental methods you devise must be accurate enough to minimize uncertainty, but precise enough for you to see the variation in your measurements so that s may be calculated with a precision of at least two significant digits.

Your team will develop and execute their experimental plan just as they would if they were involved in an engineering project in industry, with deadlines, a budget, access to a finite variety of laboratory resources, and freedom to use their ingenuity and creativity to solve the problem any way they can. Purchased items will count against the team budget, but if students can procure equipment and materials as scrap or donations, then they don't count against the budget. A set of two brass parts (a section of plate and a piece of flat bar) will be furnished to each team, along with a supply of WD-40 lubricant and metal polishing materials. The cost for all of these items will be charged against each team's budget. Other stock laboratory materials may made available to teams at no cost, depending on what equipment the lab has on hand. Any equipment students request the lab to purchase will be charged against the team's project budget.

Low values of experimental uncertainty indicate consistent experimental results, and will contribute favorably to the team member's "performance reviews," i.e., the grades on their reports. High precision in the measurement techniques will allow variation in the measurement to be measurable, while high accuracy will reduce uncertainty.

This project will extend over several weeks. You will be given a modest amount of time in class to work on the project, but you will also need to plan on spending a significant amount of time outside of class planning, acquiring materials, or working in the lab after hours on data collection.

You will be required to submit several interim reports or summaries. These do not need to be lengthy, but they do need to be typed and formally presented with a heading indicating the date, your team members, the nature of the submittal, and so on. Your final report will be a standard full lab report, written in full compliance with the standards described in *The Student Lab Report Handbook*. Your report must include an abstract, and discussion of experimental uncertainty.

Data Analysis

To expedite your data analysis, you will use a spreadsheet to calculate the mean and standard deviation for each of your four data sets. In Microsoft Excel this is simple to do. The steps are as follows:

1. Enter your data (the values of μ for one experimental configuration) into a column in the spreadsheet.

2. Click on an empty cell in the column below the data. From the tool bar select Insert and Function. In the formula builder under the *fx* tab double click on AVERAGE and hit the enter key. You will then have the mean of your data set in the spreadsheet.

3. Click on another empty cell below the data in the same column. Repeat the previous step, but this time select STDEV. A box will appear in the data column that may include the cell containing the mean, and it may or may not include your data. Click on the upper and lower right-hand corners of the box and drag the edges so the box includes only the data and not the cell containing the mean. Then hit enter. The sample standard deviation of your data will now appear in the cell.

Project Schedule

Your project milestones and critical dates are shown in the table below.

Project Milestones and Critical Dates	
Date	Activity and/or Submittal Due
	First in-class lab session. Teams begin working on experimental plan and identifying needed materials.
	Second in-class lab session. Teams continue planning and begin experimenting with possibilities for actual setup.
	Experimental plan due. This is a one-page typed summary of your plan, setup and procedure. Enough detail is required to demonstrate that you know what you are doing. Include a second page containing a complete equipment and materials list. Highlight all necessary equipment not currently on hand in the lab.
	Third in-class lab session. Teams complete setting up and begin taking data.
	Data summary due. This is a one-page typed summary of your data. Include a table showing all data for a minimum of six trials each for all four experimental setups. Include a second table showing the calculated mean and sample standard deviation values for each of the four data sets.
	Final reports due.

Project Budget

Your project budget is $_____. Expenses amounting to $_____ for the brass parts and polishing materials have already been charged against this budget, leaving your team with a balance of $_____ to complete the project.

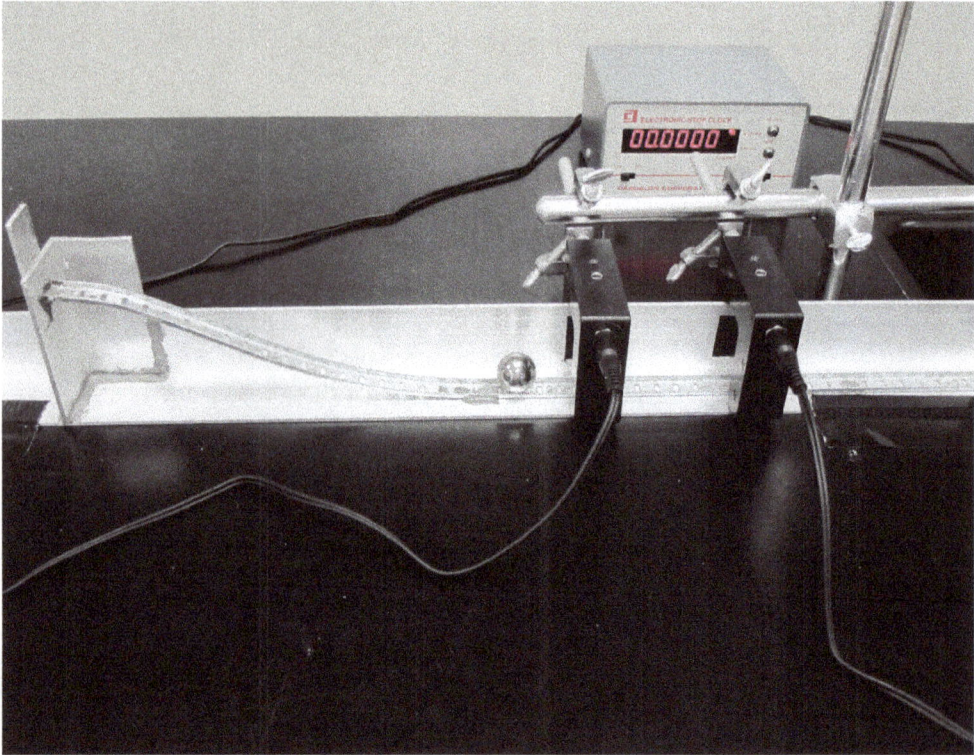

Experiment 3   Rotational Kinetic Energy

Learning Objectives

Features in this experiment support the following learning objectives:
1. General objectives for laboratory experiments (see page 4).
2. Explore and learn to use unfamiliar scientific equipment.
3. Correctly use a micrometer to make measurements.
4. Correctly use a dial caliper to make measurements.
5. Use the theory of translational and rotational kinetic energy to make quantitative predictions of experimental results.
6. Correctly format and present sophisticated mathematical development of theory.
7. Use mathematical models to determine an accurate value for the local acceleration of gravity.

This sweet little experiment is very similar to the Hot Wheels Lab presented in Part 1, Chapter 3 of this book. The same equipment is involved, and the procedures for taking data are the same. If you have not read the description of the Hot Wheels Lab, read that chapter first for a description of the general setup and equipment details. Rather than repeat all of those details here I will simply describe the particular issues that apply to this more sophisticated investigation of conservation of energy.

The conceptual difference between the two experiments is straightforward. Instead of a Hot Wheels car going down a ramp and gaining kinetic energy that is almost entirely translational, the rolling steel ball used in the present experiment picks up kinetic energy that is partly translational and partly rotational. To apply conservation principles in this context requires students to be proficient in angular (rotational) calculations of velocity and energy. Students must also be able to relate rotational and translational variables to one another.

As I wrote back in Part 1, Chapter 3, students all love the Hot Wheels Lab. But even though the students don't get to use the Hot Wheels cars in this similar experiment, the utter simplicity of the experimental design—and the numerous complications that arise, which we will discuss presently—is very interesting to them. If you can afford to use the photogates and digital timer (strongly recommended) you should be able to get the prediction-result difference ratio down to below 2.5%. Students are always intrigued when relatively simple computations based on a straightforward theory deliver predictions that can be accurately verified with such a simple apparatus.

Now, about those complications I just mentioned. We physics teachers take *great delight* in mathematical complexities, and in the challenges of deriving a mathematical, theoretical prediction. As I write this I am mentally rubbing my hands together and giggling at the mathematical gymnastics to come. Seniors in physics should be able to handle some pretty sophisticated mathematical analysis, if they are given enough information to point them in the right direction. This experiment provides just this kind of mathematical workout for them, and I don't give them much help beyond the student instructions, particularly with my students in honors-level physics classes. If they have the mathematical chops to be in senior physics in the first place, they should be able to work out the math I will present below. And you as their teacher should encourage them to *love every minute of it*. Physics *is* mathematical modeling, and mastering the math is the *sine qua non* of the discipline.

I emphasize these things here because we tend to forget that students do not normally bring this mentality with them when they come to our classes to study physics. In fact, many students think of math as something to try to avoid or minimize as much as possible, and may even hope to study physics while holding the mathematics at arm's length. But relishing the mathematical and theoretical complexities is a defining characteristic of the mentality of physicists and physics teachers. I encourage you to exhort your students on these matters often, and to seek to instill in them a love for mathematical challenge and mathematical elegance. These aspects are a profound part of the discipline of physics, and should not be absent from your classroom.

But the delight in this experiment doesn't stop at mathematics and theory. We also get very hands-on and practical by using precision machine tools in our measurements, tools students may have heard of but almost certainly have never used! This experiment combines complex theory and sophisticated tools for precision measurement! To me, this makes it just about the perfect experiment. So now that I have set the stage for all the fun we can have with this, let's dig in.

Materials Required (for the class)

1. steel ball, 1-inch diameter, such as No. AP5626 available from Flinn Scientific (flinnsci.com)

2. shelving support rail, 5/8 inch wide x 1/8 in deep, and available at hardware stores

3. stop watch (only if digital timing system is not available)

4. masking tape

5. carpenter's level

6. track support frame (described below, made of aluminum stock available at hardware stores)

7. J.B. Weld epoxy adhesive

8. stop watch

9. meter stick or machinist's rule

10. drafter's triangle

11. copy paper (for leveling)

12. micrometer

13. dial caliper

Additional optional equipment if digital timer is used:

14. digital timer, such as Daedalon Corp. model ET-41 Electronic Stop Clock

15. photogate (2), such as Daedalon Corp. model EA-27 or ET-45[1]

Additional optional mounting equipment if ET-45 photogates are used:

16. support stand (2), such as Flinn Scientific model AP5421 (flinnsci.com)

17. clamp holder (3), such as Flinn Scientific model AP8219 (flinnsci.com)

Experimental Purpose

Use the law of conservation of energy and the equations for gravitational potential energy, translational kinetic energy, and rotational kinetic energy to predict the translational velocity of a solid steel ball after it rolls to the bottom of a ramp, and compare this prediction to experimental results.

Overview

Again, the basic idea is very similar to the Hot Wheels Lab. Please read that description for a lot of background information that is not covered here.

As in the Bull's-Eye Lab (Part 3, Chapter 1), in this experiment we will use bookshelf support rail as a track for a rolling 1-inch steel ball. However, the straight track from the Bull's-Eye Lab is replaced here with an S-curve of track that will better suit our purpose. Aluminum stock (flat and angle pieces) and J.B. Weld adhesive are used to make a mounting frame for the track made of the shelving rail, illustrated in the photos beginning on page 198. Instead of the 3/8-inch deep rail used previously, here we will use 1/8-inch deep shelf mounting rail so that the track can be bent into the required S-shape. As with the Hot Wheels Lab, if you are doing your timing by hand with a stop watch, you will need about 1.5 m of track at the bottom of the hill. If you are using digital timing equipment, you only need enough horizontal track at the bottom of the hill to allow room for two photogates placed just over 10 cm apart.

On a track shaped like this the steel ball will begin its descent by sliding, causing energy to leave the system as heat produced by friction.

As you can see from the photos, the track for this experiment is shaped into an S-curve. The reason for this is as follows. If the track is shaped into a simple down ramp followed by a horizontal section (see sketch), the steel ball will begin its descent on the steel track by sliding rather than by rolling. The ball will eventually begin rolling, but the sliding at the beginning introduces a time interval in which the work done against friction is removing energy from the system, and this will reduce the accuracy of our prediction, which does not take friction into account. The simple solution is to shape the track into a S-shape, and to make sure the downward slope is not too steep so the steel ball doesn't slide.

On a track shaped like this the ball will begin its descent by rolling rather than sliding.

---

1   See "Photogate Notes" section on p. 70 for more details.

There are a number of intriguing complexities that arise in this apparently simple experimental set-up. Some of these are theoretical, and some are practical. We will look at these in detail in the following three sections.

Theoretical and Algebraic Details

The general conservation of energy equation for this experiment can be written as

$$E_{Gi} = E_{KfT} + E_{KfR}$$

where E_{Gi} is the initial gravitational potential energy of the steel ball, and E_{KfT} and E_{KfR} are the final translational and rotational kinetic energies of the rolling sphere at the bottom of the ramp. (Of course, this equation assumes that the height, h, in $E_{Gi} = mgh$ is the difference in height from the top to the bottom of the ramp.) Students will use this general conservation of energy equation to work out their predictions of the ball's translational velocity at the bottom of the hill. Students should be expected to insert the formulas for the various terms in the general conservation equation and express the predicted velocity in simplified form, both in terms of the standard variables such as the radius of the sphere, as well as in the actual measured variables, such as the diameter of the sphere.

These algebraic manipulations get rather interesting due to the fact that, unlike a sphere rolling on a flat hard surface (like a table top), for a sphere rolling on a track the radius of the sphere that appears in the moment of inertia ($I = \frac{2}{5}mr^2$) is not the same as the rolling radius of the sphere, which I will designate as R. The accompanying diagram locates these variables with respect to physical geometry of the sphere on the track. Also shown is the

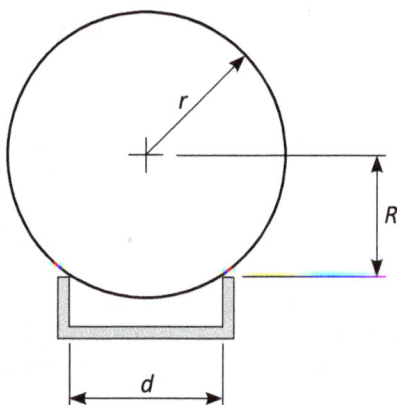

Cross-sectional view of the sphere on the track.

width of the inner groove of the track, d.

Now, for a ball rolling on a table top (that is, $r = R$), it is not difficult to show that the velocity of the ball after falling through a height h equals

$$v = \sqrt{\tfrac{10}{7} gh}$$

However, on a track with width d the velocity will be

$$v = \sqrt{\dfrac{gh}{\frac{1}{2} + \dfrac{r^2}{5\left(r^2 - \frac{d^2}{4}\right)}}}$$

In the actual experiment we will measure d directly. But we cannot conveniently measure r directly. Instead, we will measure the diameter of the sphere, D. So re-expressing the predicted velocity in terms of variable we can measure directly gives

$$v = \sqrt{\dfrac{gh}{\frac{1}{2} + \dfrac{D^2}{5\left(D^2 - d^2\right)}}}$$

Finally, expressing this in a more elegant, simplified form we have

$$v = \sqrt{gh\dfrac{10\left(D^2 - d^2\right)}{7D^2 - 5d^2}}$$

Notice from this expression that as the track width, d, reduces to zero the predicted velocity approaches the simple expression for a ball rolling on a table top.

None of the equations shown here should be given to the students. They should work them all out themselves after the class has studied rotational kinetic energy and moment of inertia and their application to the law of conservation of energy. Students should work these computations out prior to the day of the experiment. If they do, then they will come to the experiment knowing that measuring the mass of the ball is unnecessary.

Refining the Value for g for Improved Accuracy

If you are using a digital timer for this experiment, then this experiment becomes an opportunity for students to enjoy working with high accuracy and high precision, resulting in a very low prediction-result difference ratio. You have an opportunity to increase the accuracy even further by refining the value used for g, the acceleration due to gravity on the surface of the earth. With a digital timer and the instruments described in the next section, students will be able to make every measurement to at least four digits of precision. Thus, it will be appropriate to discuss methods for obtaining a value for g at your location that is more precise than 9.8 m/s² and more accurate than the global average of 9.80 or 9.81 m/s² typically found in textbooks. In 2012 I used the methods below with the data my class took, and our 2.48% difference ratio was reduced to 2.44%. This is only a very small improvement, but when it comes to reducing the difference between prediction and experimental result I am interested in any improvement that will enhance the class (i.e., bring in more physics) without requiring undue time and expense.

Engaging in this discussion with students is bound to be interesting. The earth is not a perfect sphere; according to the reference listed below, the diameter at the equator is approximately 43 km larger than the diameter at the poles, and thus the value of g depends on one's latitude. Moreover, since gravitational attraction follows an inverse square law with respect to the distance between the centers of the object, the value of g also depends on one's altitude.

There are several ways to obtain a more precise value for g. One way would be to devise an experiment that could measure g at your location to four significant digits. This would not be unreasonably difficult and would be a lot of fun. However, I am not including that experiment in this book, so you are on your own.

A second method is to find a value for g at your location in a reference resource. Eugene Hecht's great physics text *Physics: Alegbra/Trig* has a small table listing values for 15 different cities and locations. Incredibly, my town, Austin, Texas is listed there, and the value given for g is 9.793 m/s². There are probably on-line resources tabulating values for various locations, if you happen to live at a location popular enough to be listed.

A third possibility is to find a model for calculating g, based on your latitude and altitude (which may easily be determined through various online resources[2]). John Rickert, Associate Professor of Mathematics at Rose-Hulman Institute of Technology, has a site with a model for calculating g this way.[3] At a latitude of L (in radians), Professor Rickert's model states that the value for g (to seven digits of precision!) will be approximately

$$g = 9.780327 \left[1 + 0.0053024 \sin^2(L) - 0.0000058 \sin^2(2L) \right] \text{ m/s}^2$$

After computing g based on your latitude, you can then make another small correction for your altitude. Professor Rickert models the earth as an ellipsoid, with an equatorial radius of approximately 6,378,140 meters and a polar radius of approximately 6,356,755 meters. This model gives a value for the local acceleration of gravity, a, as a function of your previously calculated value for g, as

$$a = g \frac{R^2}{\left(R + H \right)^2}$$

where H is the altitude above sea level in meters, and R is the radius of the earth at your location, which is approximately

$$R = 6,356,755 \sqrt{1 + 0.0067396 \cos^2(L)}$$

Now, it may seem way over the top to go to all this effort to model g for a reduction in the prediction-result difference ratio of only 0.04%. Perhaps. But a couple of subtle points may be made here that can really enrich your class. First, as I wrote at the beginning of this chapter, physics is all about mathematical modeling. Most of our students probably think that 9.8 m/s² is some kind of magic number. They typically have no idea that it varies according to both latitude and altitude, and deriving a mathematical model for it has probably never even occurred to them. So just raising the issue has a lot of instructional value for students at this level. Second, the added mathematical complexity is not all that difficult to incorporate, and the discussion surrounding it is a lot of fun. (Deriving Professor Rickert's model itself would

---

2   See, for example, altitude.org/find_altitude.php and findlatitudeandlongitude.com.

3   rose-hulman.edu/~rickert/Classes/ma112/gravity.html

be difficult, but we can just use the model he has worked out.) Finally, it is *really* fun to be a bloodhound for finding ways to reduce the difference between prediction and experimental result. I like communicating to my students that when we find a way to reduce the difference by 0.04% with only a few minutes of calculator work, it *is* worth it, and we get excited about it as we reach for our calculators!

By the way, the value for *g* at Austin, Texas given in Hecht's text is 9.793 m/s². For the location of the lab where we performed this experiment, the value for *g* I calculated from Rickert's model for our latitude is 9.793 m/s² (same as the Hecht value), and 9.792 m/s² when our altitude (not great, only 244.6 m) is taken into account.

Precision Measurements

Of the various length measurements in this experiment, three are somewhat challenging: the change in the height of the ball, the width of the groove in the track, and the diameter of the ball. A few comments about each of these are warranted.

The difficulty in measuring the height of the ball is in avoiding parallax error. Locating the center of the ball with accuracy is not at all convenient, so one has to use the top of the ball for measuring the change in height. And since the ball is spherical, the top of it is going to be about 1.25 cm away from the rule being used for the measurement. I solve this problem by using a drafting triangle to locate the top of the ball, holding the perpendicular edge of the triangle parallel to the rule, as shown in one of the photos.

For the groove width and ball diameter we are going to use a dial caliper and a micrometer, machine tools familiar to any machinist or hands-on engineer but often not so familiar to laymen. These precision instruments are *fascinating*. Your students will stand *transfixed* as you explain how to operate these devices, and they won't be able to wait to try them out for themselves. This is one of those times when bringing in genuine, precision tools for students to work with (instead of mediocre made-for-high-school stuff) pays dividends in generating student interest and in teaching them valuable skills that can come in handy later if they enter a technical field involving precision measurement.

The groove width can be measured with a rule if necessary. But a more precise way to measure it is with a dial caliper. As shown in the photo, this device can measure an inside dimension and return a value to the nearest 0.0001 inch. This level of precision is not possible with a rule. Students should be challenged to record the groove width at multiple locations of the track and work out a method of averaging the measurements. Some kind of weighted average will probably be necessary since the groove width is likely to be different along the curves from what it is on the straights because of deformation of the rail during bending. Inexpensive dial calipers are available from discount tool suppliers in the range of $30-50. If you are going to invest in one for a school science lab, you should consider saving up the $200 necessary to buy a good one.

Finally, the only tool that can measure the diameter of the ball with any accuracy is a micrometer. A good micrometer can measure directly across the diameter of the ball, and a good quality instrument can deliver resolution to the nearest 0.0001 inch.

These days high quality calipers and micrometers all come with digital indicators. These are nice, but don't depend on the digital indicator. Make sure the instrument you purchase also has the markings necessary to read the measurement manually. Teaching the students to read these instruments is a big part of the fun of the experiment.

Learning how to handle the dial caliper and micrometer, and how to read the measurements from them, takes some patience. If you would like to see a video tutorial on the use of these instruments, please visit the video tutorials page on our website at novarescienceandmath.com.

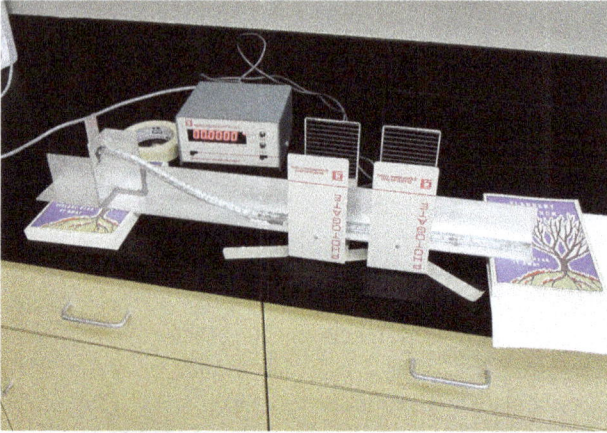

This photo shows the complete setup using the EA-27 photogates, which have large flanges to allow them to be fastened to an air track. Books and copy paper are used to level the track and to situate the track over the photogates.

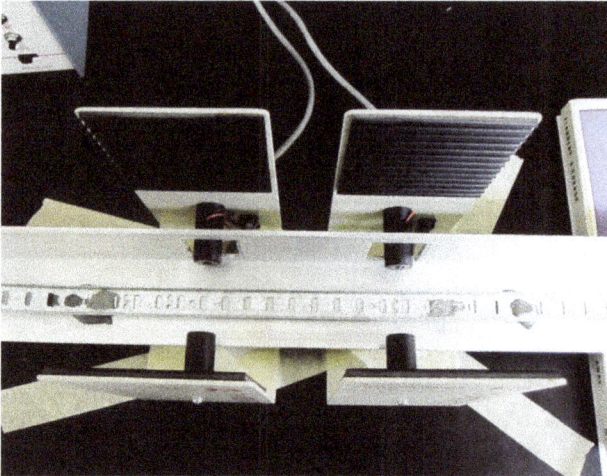

A close-up of the EA-27 photogates. The black sensors use infrared light to trigger the timer.

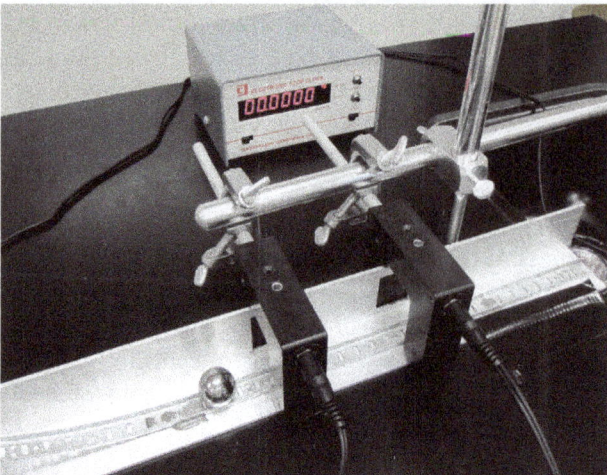

A close-up of the timer assembly using the ET-45 photogates.

A close-up of the top of the ramp. The gentle initial slope helps get the ball rolling without initial sliding of the ball. The support frame is cut from aluminum stock and glued together with J.B. Weld epoxy adhesive. The design of the frame provides a reliable backstop so the ball consistently begins from the same height.

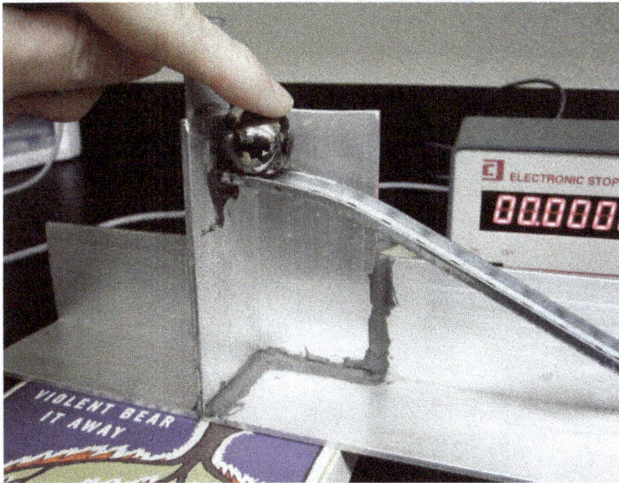

Ready to roll!

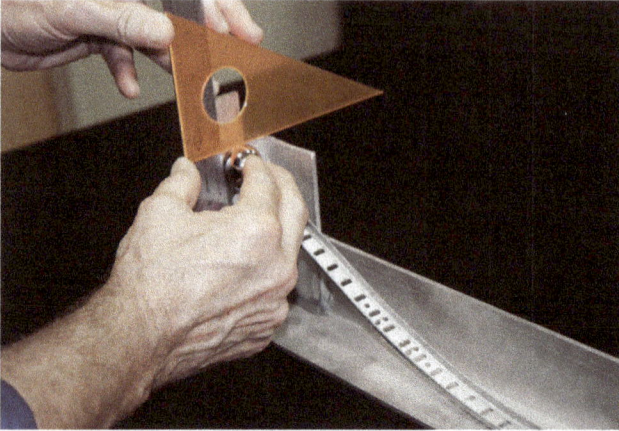

Technique for measuring the height of the ball. In this photo I am holding all the items myself, which is rather cumbersome, and the position of my right hand and thumb doesn't really even allow the measurement to be taken. If students work together on this, the lower edge of the triangle can effectively be used to read the height of the top of the ball, as long as the edge of the triangle is aligned with the edge of the rule.

Measuring the width of the track with a dial caliper.

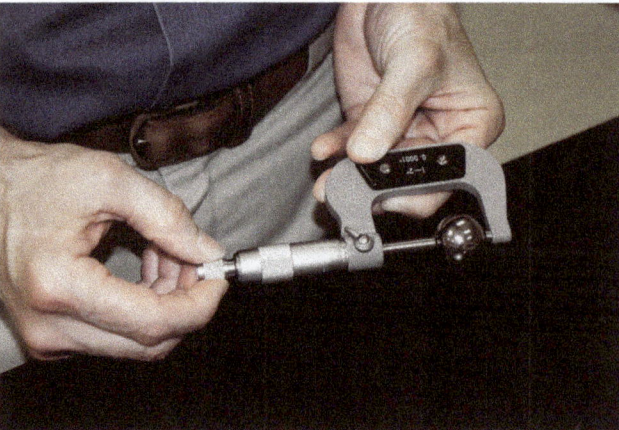

Measuring the diameter of the ball with a micrometer.

Rotational Kinetic Energy
Applying Conservation of Energy to Rotating Objects

In this experiment a steel ball will be stationed on a track at the top of a ramp. We will release the ball and allow it to roll down the ramp, converting gravitational potential energy into kinetic energy (translational and rotational). Your goal is to use the principles of conservation of energy applied to rotating and translating objects to predict the translational velocity, v, of the ball when it reaches the level part of the track at the bottom of the ramp, and compare this prediction to the actual velocity determined by distance and time measurements. Your hypothesis is that the principles of conservation of energy can be used to give an accurate prediction of the ball's final translational velocity, v.

Measurements

The setup for this experiment is relatively simple, but to enable the highest possible accuracy we will be using some sophisticated measurement techniques and instruments.

Several measurements are involved in forming your prediction, each of which should be made multiple times by different participants. To work out the predicted velocity from energy principles requires that you measure the change in height the ball will undergo. Relating together the rotational and translational motion of the ball requires that you measure the diameter of the ball and the width of the track (see next section). Your instructor will review these measurements with you, and perhaps demonstrate for you how to make high-precision measurements using a dial caliper and micrometer, which are needed for high-precision measurements of track width and ball diameter.

There are also two measurements associated with determining the experimental value of the ball's velocity. You will need to time the ball on the level part of the track as it passes through a timing zone of known width. This may be done with a stop watch, if necessary, in a timing zone about 150 cm long. Preferably, use a digital timing system for this measurement to improve the accuracy of your results. With a digital timing system you can reduce the width of the timing zone to just a few centimeters. To reduce losses from friction you will want your timing zone to be as short as possible, but the precision of your measurement of the timing zone length, which you will measure with a rule, will gain a significant digit if the timing zone is just over 10 cm long, instead of being less than 10 cm long.

Before the Experiment Starts

The figure on the next page represents the ball sitting on the track as it will be in the experiment. The inside width of the track on which the ball sits is d, and the radius of the ball is r. Notice that although the radius of the ball is r, the "rolling radius" of the ball is R, not r. In other words, the ball is not rolling on a surface at its full radius, but is sitting down in the track and is rolling on the smaller radius, R. Thus, when using the radius of the ball in an expression pertaining to the moment of inertia of the ball, you

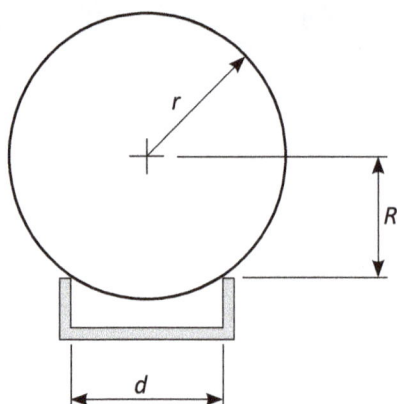

Cross-sectional view of the ball on the track.

must use the ball's actual radius, r. But when relating the translational velocity v to the rotational velocity ω, you need to use R.

Before the experiment, write a conservation of energy equation in which initial gravitational potential energy is converted into translational and rotational kinetic energy. Eliminate the variable ω in favor of v and R. Your equation will then contain only the variables h, v, r, R and d. R is not easy to measure, so use the geometry apparent in the diagram to eliminate R by replacing it with an expression based on r and d. The ball radius, r, is not easy to measure either, although measuring the diameter of the ball, D, is straightforward with a micrometer. So eliminate r from your equation too, replacing it with $D/2$.

Finally, solve your equation for v, so that you have an equation expressing your predicted velocity v in terms of variables that may all be directly measured. To make this equation easy to use, simplify your expression for v so that it is as simple and elegant as possible.

Refinements in the Acceleration Due to Gravity, *g*

If you are using a digital timing system your accuracy in this experiment should be very good (that is, low prediction-result difference ratio). To improve your accuracy even further, consider refining the value you use for g to take into account the fact that the value of g varies with both latitude and altitude. There are various options for obtaining a value for g that is more precise than the 9.80 m/s² we typically use (which is really just a global average, and applies better in the northern part of the U.S. than it does in the south, because the earth is not perfectly spherical). Your instructor will review these options with you.

Your Report

After forming your prediction and taking data, prepare a complete report outlining your theory, calculations, data, prediction-result difference ratio and analysis. In your theoretical background you should include a presentation of the major equations and steps involved in the algebraic analysis described above. In this presentation you do not need to show every single algebraic step (like an introductory algebra book would), but you do need to show enough steps in your derivation to make it clear that you understand the equations. Proper style requires that you talk through your steps as you present them, as textbooks do. You cannot simply write equations one after another down the page the way one does when solving a problem.

High accuracy and precision are a major objective for this experiment. Accordingly, be sure to use significant digits correctly in all of the measurements, data reporting and calculations.

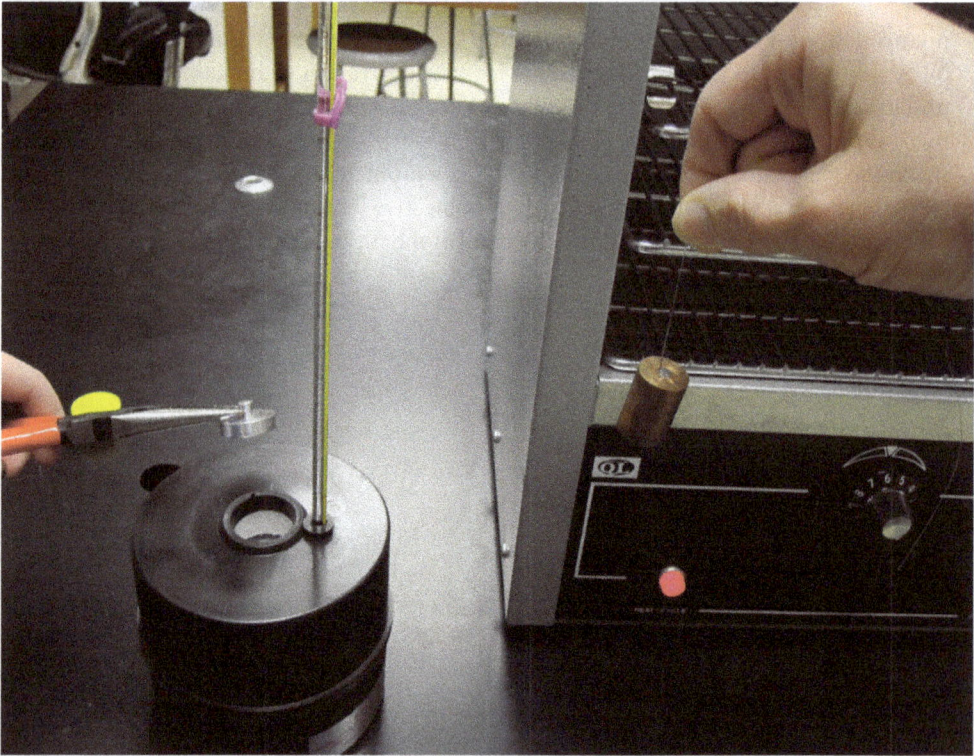

Experiment 4 Calorimetry

Learning Objectives

Features in this experiment support the following learning objectives:
1. General objectives for laboratory experiments (see page 4).
2. Explore and learn to use unfamiliar scientific equipment.
3. Use the theory of calorimetry and heat transfer to make quantitative predictions of experimental results.
4. Correctly format and present sophisticated mathematical development of theory.
5. Skillfully handle sensitive laboratory apparatus.

A common exercise in the study of specific heat capacity and calorimetry is to combine two materials at different starting temperatures and calculate the resulting final temperature. A variant of this standard problem applies to the case of two materials, one for which the specific heat capacity is known, and another for which the specific heat capacity is unknown. In this case the initial and final temperatures and the known specific heat capacity are used to determine the unknown specific heat capacity. This is the approach taken in this investigation.

As with several of the experiments in Parts 1 and 3 of this book, this experiment lends itself to the class working together as a single group. This is because the equipment needed to produce hot or cold environments without the use of water—refrigerators and ovens—is typically large and expensive, so having the class work together as one group makes sense. In the absence of this equipment one can use ice water baths and boiling water to provide the hot and cold temperatures needed. However, any time water is involved accuracy will be compromised. Water's high specific heat capacity means that any water that clings to the sample under test (usually a metal) will have a noticeable effect on the results, and trying to dry the sample before placing it in the calorimeter means that the sample is already cooling or warming due to the room temperature environment, which also seriously compromises accuracy. Thus, I have designed this experiment to keep everything dry in order to provide the best accuracy in the results.

In this experiment we will treat a copper sample as a material with an unknown specific heat capacity. We will use the techniques of calorimetry to determine the specific heat capacity of the copper sample, and compare this experimental value to values found in standard references.

Materials Required (for the class)

1. aneroid dry calorimeter (2), such as Flinn No. AP9160 (available from Flinn Scientific, flinnsci.com)
2. specific heat cylinders (2), such as Flinn No. AP6112 (available from Flinn Scientific, flinnsci.com)

3. thermometer (3)

4. needle-nose pliers

5. laboratory oven

6. freezer

7. fishing line

8. laboratory balance

9. J.B. Weld epoxy adhesive

10. thermocouple with digital reader, such as PASCO Xplorer GLX with Type K temperature sensor PS-2134 (available from PASCO at pasco.com), or Fluke model 179 digital multimeter with model 80BK temperature probe (available from Allied Electronics at alliedelec.com)

Experimental Purpose

Experimentally determine the specific heat capacity of copper using the techniques of calorimetry, and compare the experimental value to the standard reference value.

Overview

The aneroid dry calorimeter used for this experiment consists of a cylindrical aluminum core encased in a layer of insulating Styrofoam, all enclosed in a protective plastic casing. We will treat the specific heat capacity of the aluminum cylinder as a known quantity. The aluminum cylinder has a cylindrical chamber bored in it with an aluminum lid, and this chamber is sized to accept the copper specific heat capacity cylinder. The calorimeter also has a hole drilled in it for a standard glass thermometer.

The basic plan is to place the copper sample in a cold or hot environment while keeping the calorimeter at room temperature. After measuring and recording the initial temperatures of the copper sample and the calorimeter core, the copper sample is removed from its initial environment, quickly placed inside the calorimeter, and covered with the aluminum and plastic lids. The thermometer in the calorimeter is monitored until thermal equilibrium is reached and the final temperature is recorded. Waiting for the copper sample and calorimeter to reach thermal equilibrium requires about 6-8 minutes. Finally, the masses of the aluminum calorimeter core and the copper sample are measured. With the three initial and final temperature values, the two metal masses, and the specific heat capacity of the aluminum, the specific heat capacity of the copper sample may be calculated by assuming that the heat gained by one metal is equal to the negative of the heat lost by the other metal. In terms of the variables involved,

$$c_1 m_1 \Delta T_1 = -c_2 m_2 \Delta T_2$$

or,

$$c_2 = -\frac{c_1 m_1 \Delta T_1}{m_2 \Delta T_2}$$

where ΔT is defined as $T_f - T_i$.

The entire experiment is performed twice, once with a copper sample in a freezer at approximately –15°C, and once with a copper sample in an oven at approximately 120°C. This will allow two different experimental values of the copper specific heat capacity to be calculated and compared. One could do this with a single calorimeter and copper sample, but this would require waiting between trials for the aluminum core of the calorimeter to return to room temperature, which would probably mean doing the two trials on different days. I prefer to set up the hot and cold trials simultaneously, placing the copper cylinders and the temperature sensors inside the hot oven and cold freezer at least two hours prior to the experiment. This time interval should be adequate to assure that the oven and freezer temperatures are as stable as possible, and that the copper cylinders are both at thermal equilibrium with their environments.

Temperature Measurements

The initial and final temperatures of the calorimeters are read with the thermometers inserted into the calorimeters. The accuracy of these measurements is an issue which I will address in the Pre-Lab and Post-Lab sections a bit later. But accuracy aside, making these measurements is simple and inexpensive. Measuring the temperatures in the freezer and oven can be a lot trickier. The bottom line is that you must have a way of measuring the temperature before opening the door, because the air temperature in a freezer or oven changes instantly when the door is opened. Further, the air temperature inside a freezer can vary by 8 or 10 degrees as the freezer refrigeration system cycles on and off. Thus, simply measuring the air temperature inside the freezer will not give the temperature of the copper cylinder itself.

A laboratory oven should have a hole in the top for installing a thermometer to monitor the oven temperature. Using a thermometer here again raises the accuracy issue we will get to later, but this is definitely the cheapest and most convenient way to measure the oven temperature. From my own experience, it seems that the temperature in a laboratory oven is regulated much more tightly than the temperature in a consumer-grade freezer, so a thermometer in the oven will give a reasonably accurate reading of the temperature of the copper cylinder. All things considered, I am willing to use the thermometer for the oven.

The temperature of the copper cylinder in the freezer is a different matter. In addition to the temperature fluctuations mentioned above, there is no convenient way to use a thermometer to acquire the temperature inside without opening the door. I propose solving this problem by using a thermocouple to measure the temperature *inside* the copper sample while the sample is the freezer. The digital reader or multimeter can be outside the freezer, the thermocouple extended inside, and the door closed on the thermocouple lead, which is small enough not to cause a problem with air leakage. To deal with the problem of temperature cycling, we drill a hole in the copper sample and insert the thermocouple into the hole, as described in the next section.

As with many of the other experiments and demonstrations in this book, this experiment may be performed inexpensively, using an ice water bath and boiling water as initial environments for the copper cylinder in place of the freezer and oven. However, as mentioned previously, it is difficult to dry the water off the copper cylinder quickly enough to achieve the best accuracy.

Aluminum and Copper Details

According to the manufacturer, the aneroid dry calorimeter sold by Flinn Scientific as their item no. AP9160 has an aluminum core made of 6061 aluminum alloy. According to a host of online references[1], the specific heat capacity of this alloy is 0.896 J/g·°C. This value is treated as a known quantity and is given to the students.

I will now address two modifications to the copper cylinders that are in order, both of which will improve the accuracy of this experiment.

The first modification is to drill a small hole in the side of the cold-trial copper cylinder about 3/8 inch deep. The thermocouple or temperature sensor lead is inserted into this hole, providing a direct measurement of the temperature inside the copper cylinder. You want to size the hole for your particular sensor so that the temperature sensor lead fits in with just a small bit of friction. This will help assure that the lead remains firmly situated inside the cylinder and does not slip out.

The second modification is to the hook on the top of the copper cylinder. The particular cylinders supplied by Flinn, which are conveniently sized to fit the chamber in the calorimeter, have an ordinary cup hook inserted in the top. Unfortunately, this hook is not made of copper. It is an unknown plated metal, meaning that it has a different specific heat capacity than copper and thus must be removed. (Years ago some science suppliers stocked copper cylinders made entirely of copper and without the cup hook. It seems that as of the writing of this book the suppliers all supply the same style of cylinder, with the hook, even when the pictures in the catalog look otherwise! So we will just deal with it.)

Copper cylinders with the cups hooks removed and replaced with fishing line held in place with a small quantity of J.B. Weld epoxy. The cylinder on the right has the hole for the insertion of the thermocouple lead for the cold trial.

The cup hook will screw out of the cylinder with a pair of pliers if you heat the copper cylinder up to make the metal expand. If you try to screw out the hook with the copper at room temperature the hook may shear off, and then the only way to get the embedded screw out is to drill it out (not easy to do).

After removing the hook, mix up some J.B. Weld epoxy adhesive and cut off a piece of fishing line about 12 inches long. Insert the fishing line down into the hole where the cup hook was and fill the hole with a couple of drops of J.B. Weld, wiping off any excess epoxy that gets on the cylinder in the process.

This modification accomplishes two things. First, the fishing line may be grasped directly with the hand, and thus provides a convenient way for the student to lift the cylinder out of the oven or freezer, swing it over to the calorimeter, and lower it down into the calorimeter chamber. Second, removing the cup hook should provide

---

1   E.g., http://asm.matweb.com/search/SpecificMaterial.asp?bassnum=MA6061O

a small improvement in accuracy, although the improvement is not as great as one might initially think. Even though the hook is not copper, it is at least a metal, and thus possesses a specific heat capacity in the same neighborhood as copper. The epoxy, by contrast, is an insulating plastic compound with a specific heat capacity 3-4 times higher than that of copper. However, a couple of drops of the epoxy placed in the hole will have only a small fraction of the mass of the hook that was removed. Inserting a foreign material with three times the specific heat capacity but only, say, a tenth the mass represents a trade-off in the right direction.

Pre-Lab Discussion

Review the following details with students prior to the experiment.

1. Explain what a calorimeter is. It is helpful to compare scientific calorimeters to other insulating containers students are already familiar with, such as Thermos bottles. Discuss the benefits of using a dry calorimeter (no water) when doing calorimetry experiments. These benefits are due to the very high specific heat capacity of water compared to just about everything else, especially metals.

2. Experimentally determining the specific heat capacity of copper is the goal, but we also want to know how accurate our measurement is. This is why copper was chosen as the material with the unknown value. The value actually is known and is easy to look up in standard references. For calculating the prediction-result difference ratio, we will use the standard value as the predicted value and the result of the experiment as the experimental value.

3. When transferring the copper cylinder from the oven or freezer to the calorimeter students must move quickly but carefully. The goal is to minimize the time the cylinder is exposed to the room temperature air. Station the calorimeter as close to the oven or freezer door as possible. Have one student poised holding the aluminum calorimeter lid with a pair of needle-nose pliers. Another student will remove the copper cylinder from the oven or freezer, holding it by the fishing line, and carefully but quickly place the copper cylinder into the calorimeter. Then the student holding the aluminum lid quickly places the lid on, and then puts in the plastic cap as an insulator. To minimize the possibility of something going wrong, it will be best for students to rehearse these maneuvers a day in advance.

4. Students should expect the experimental result from the hot trial to be more accurate than the value from the cold trial. Students should try to figure out why this is the case. If questions remain, the issue can be discussed further after the experiment.

Post-Lab Discussion

Revisit the issue of the prediction-result difference ratios resulting from the hot and cold trials. If, as expected, the hot trial is more accurate, ask students to discuss and explain why. As clues, ask the students about the precision and accuracy inherent in each of the measurements they took. Which device has the lowest accuracy? Students probably do not know that the accuracy of a typical laboratory thermometer (used for measuring T_i and T_f for the aluminum) is $\pm 1°C$, but once you tell them this they may begin to see where the issue lies. Next, ask students to consider what can be done to minimize the effect of the low thermometer accuracy. After some work, some students will begin to realize that since the results depend on a *change* in temperature, using a higher change in temperature will mean that inaccuracies in one of the temperature readings will have less of an effect on the overall result. In other words, inaccuracy in ΔT caused by variation in T_f will be reduced if the variation in T_f (which is a degree or two) is a lower percentage of ΔT. Ask students to cite the aluminum ΔT values for each of the trials. (The ΔT for the cold trial will be in the ballpark of 10°C. For the hot trial the ΔT should be at least twice as high.) Students should include analysis of this issue in the Discussion section in their own lab reports.

Student Instructions

A set of instructions you may reproduce and give to students begins after the following illustrations.

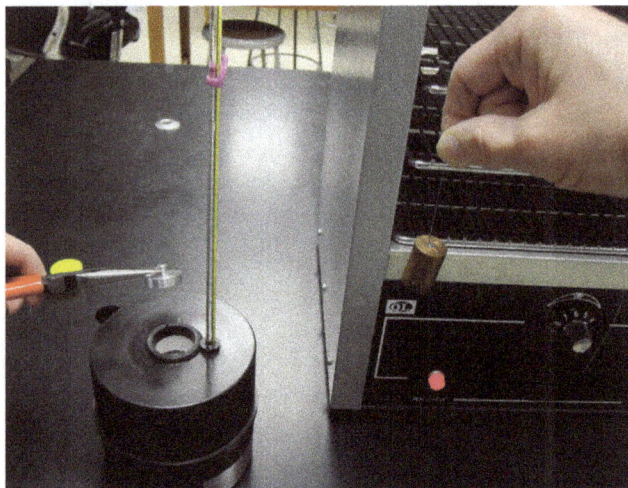

Transferring the hot copper cylinder to the calorimeter. Another student is ready to put on the aluminum lid as soon as the copper cylinder is in place.

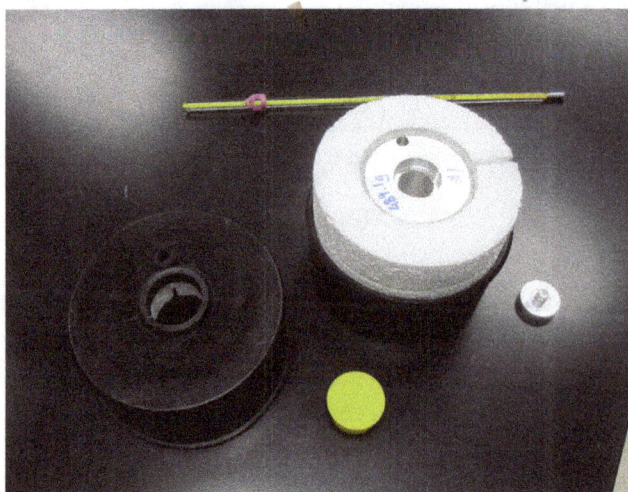

The calorimeter parts and thermometer. I decided that removing the aluminum core from the Styrofoam jacket every year to get the aluminum mass was eventually going to tear up the Styrofoam. So I wrote the mass on the core with a permanent marker to save removing it to measure its mass. The two calorimeters have slightly different core masses, so I numbered them #1 and #2 and wrote these numbers on the core and lid of each calorimeter.

The thermocouple is inserted into a hole in the copper cylinder for an effective measurement of the metal temperature. In this photo you can see that I bent the end of the thermocouple so that it would press against the inside of the hole and stay fixed in there instead of slipping out when I closed the freezer door.

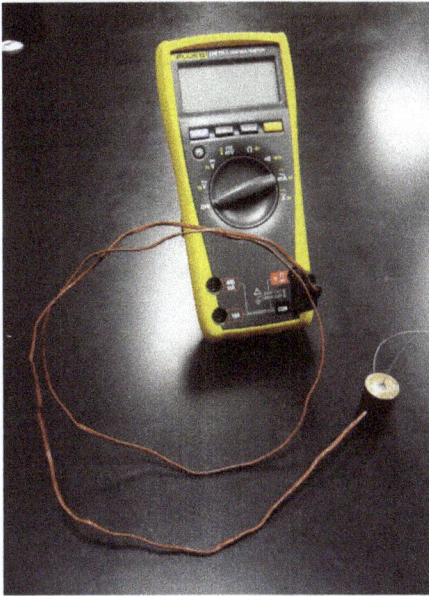

Our Fluke DMM with the thermocouple and copper cylinder.

Our test setup at the freezer. Before opening the door we take our temperature reading. Then the instant the student opens the door, the student picks up the cylinder by the fishing line, pulls out the thermocouple, and places the cylinder in the nearby calorimeter. Another student is holding the calorimeter lid with a pair of needle-nose pliers, ready to put it on as soon as the copper cylinder is in place.

Calorimetry
Experimentally Determining Specific Heat Capacity

In this experiment you will determine the specific heat capacity of a metal specimen, comparing the experimental value to a standard value for that metal to determine the prediction-result difference ratio. We will use a small copper test cylinder as the "unknown material." You will go through the experimental procedure twice, once with a hot specimen, in which the copper cylinder has been heated in an oven to approximately 120°C, and once with a cold specimen, in which the copper cylinder has been chilled in a freezer to approximately −15°C.

The experimental procedure you will use involves placing the hot or cold copper specimen in an aluminum aneroid dry calorimeter, and then using the mass, temperature, and specific heat capacity data in a calorimetry calculation to determine the "unknown" specific heat capacity of the specimen. When the specimen is placed into the calorimeter we have two masses of material (two different metals) that are not in thermal equilibrium with each other. The insulation of the calorimeter isolates the system from the environment of the room. The initial temperatures of the copper cylinder and aluminum calorimeter core, final equilibrium temperature, metal masses, and specific heat capacity of the aluminum calorimeter core may be used to determine the specific heat capacity of the specimen.

Procedural Outline

The steps below apply to the cold trial. The hot trial is performed similarly.

1. Read and record the temperature of the thermometer in the calorimeter, which is in thermal equilibrium with the classroom or laboratory. This is the initial temperature of the aluminum.

2. Read and record the temperature of the thermocouple in the freezer, which is the initial temperature of the specimen.

3. Very quickly and very carefully transfer the specimen to the calorimeter. Close the calorimeter and begin temperature data collection. Record the calorimeter temperature every minute. When four identical readings have been obtained you may assume thermal equilibrium has been achieved.

4. Record the masses of the aluminum calorimeter core (including the aluminum lid) and the copper specimen.

Analysis

To determine your prediction-result difference ratio, compare your two experimental values for the specific heat capacity of copper to a value found in a standard scientific or engineering reference. One such reference is *Marks' Standard Reference Handbook for Mechanical Engineers.*

From the design of this experiment, it is apparent (but not obvious) that we should expect the prediction-result difference ratio for the hot trial to be lower than the difference ratio for the cold trial. You should consider why this is the case, seek to understand it, and discuss it in the Discussion section of your report.

The *Marks' Handbook* quotes specific heat values in units of $\frac{BTU}{lb \cdot {}^\circ F}$. Converting these units into J/g·°C may require a bit of thought. Here are some tips: For converting pounds to kilograms, we should always proceed by converting to newtons first and then using the weight equation with an appropriate value for g to determine the mass in kilograms. This will help you to figure out the lb conversion.

For converting the temperature units from °F to °C, note that in this conversion you do not need to worry about the 32° shift in the freezing point on the Fahrenheit scale. This is because the units of measure for specific heat capacity are simply a ratio of energy per unit mass and per degree of temperature *change*; the absolute temperature is not at issue. Thus, the only thing that matters is the relative sizes of one degree on the Celsius and Fahrenheit scales. As is clear from the standard conversion equation, a temperature change of 1°F is only 5/9 as great as a temperature change of 1°C.

Reference Data

The aluminum core of the calorimeter is made of 6061 aluminum alloy. According to the numerous online references[1], the specific heat capacity for this alloy is 0.896 J/g·°C. Use this value in your calorimetry calculations.

---

1   E.g., http://asm.matweb.com/search/SpecificMaterial.asp?bassnum=MA6061O

Experiment 5   Sound

Learning Objectives

Features in this experiment support the following learning objectives:
1. General objectives for laboratory experiments (see page 4).
2. Explore and learn to use unfamiliar scientific equipment.
3. Use the theory of inverse square variation and logarithms to make quantitative predictions of experimental results.
4. Correctly format and present sophisticated mathematical development of theory.
5. Skillfully handle sensitive laboratory apparatus.
6. Use computer tools to calculate the standard deviation as an estimate of uncertainty, and represent uncertainty with error bars on a graph.

Here is yet another fun experiment that can be performed with the entire class and for very little cost. Students always enjoy this experiment—it is performed outdoors, which is where they would rather be anyway on a nice spring day!

The concept here is simple. We hook up a battery-powered piezo siren and measure the sound pressure level (SPL) in decibels (dB) at various distances, starting at one meter in front of the siren and going up to 20 m. These are the experimental data. Then we use the inverse square law for sound intensity levels and the SPL measurement at one meter to develop a set of predicted values for the SPL at the same distances where the measurements were taken.[1] Students then plot the two sets of values (actual and predicted) on the same pair of axes for comparison and analysis. As with the Friction Challenge (see Part 3, Chapter 2), students will gain additional familiarity with computer tools by using a spreadsheet to calculate the mean and sample standard deviation for the SPL measurements at each location.

Material (for the class)

1. sound level meter, such as Radio Shack Model Digital 2055, Cat. No. 33-099 (available at radioshack.com)

2. piezo siren, such as Radio Shack Model 273-079 (available at radioshack.com)

3. measuring tape, such as 30 meter, wind-up metric tape AP6323 available from Flinn Scientific (flinnsci.com)

4. lantern battery, 6V

5. duct tape

---

1   Sound intensity level and sound pressure level are defined differently, but are numerically equivalent.

6. copper tubing, 1/2 inch x 6 feet long (Available in the plumbing department of hardware stores. Any other lightweight pole or pipe will work just as well.)

7. alligator clip test lead, 36-inch (2), such as Allied Stock No. 70209822 (this stock number includes a pair of leads, one red and one black), available from Allied Electronics (alliedelec.com)

8. folding step stool

9. disposable ear plugs (one pair per person), available at large drug stores. Also available in bulk, industrial style as Grainger Item No. 6XF60 (grainger.com).

Overview

The trickiest part of this experiment is finding a suitable outdoor location for it. The ideal location is in a large, quiet field, such as a football or soccer field, far away from reflective surfaces such as buildings, and remote enough from busy roads so that traffic is inaudible. You want a time when no lawn mowers or other machinery can be heard in the area. Ideally, you also want your location to be free of tweeting birds and rumbling aircraft. (This may sound funny, but at our school there is one field surrounded by woods with birds constantly singing and chirping, and another field that the birds don't care much for. The quiet one is our spot.) Meeting all of these conditions perfectly is virtually impossible, but if you have a large field that is reasonably quiet the experiment will work. The sound level meter listed in the materials list comes with a wind screen, which will help a lot. The wind screen should always be used when taking readings outdoors.

After you have identified a usable location, you need to hope for a windless day, not very common in the spring when this experiment would typically be performed. The experiment uses an instrument that responds to the air pressure fluctuations we call sound waves. Wind also happens to involve large fluctuations in air pressure, which will make the instrument give false readings, so avoid wind if possible.

If windless conditions are not available you will at least need to wait for a day that is free from gusting wind. A gentle breeze we can handle; a blustery day (alas, typical for spring) will be hopeless.

The setup is easy. As illustrated by the photo at the beginning of the chapter, we set up a folding step stool to hold the heavy 6-V battery. The photo shows two batteries, but one 6-V lantern battery will usually work. What won't work is the 9-V "transistor" batteries. These have the right voltage, but can only drive the current required by the siren for a minute or two before the battery expires and the siren shuts off. Stick with the big 6-V lantern batteries.

The stool is also used as a support for the siren mounting pole (the copper pipe). Tape the copper pipe to the step stool to hold it vertical. Tape the piezo siren to the pole about five feet above the ground, or as high as the alligator clip leads will reach. Then stretch out the meter tape for a distance of 20+ m along the ground. As shown in the photo, it is convenient to hook the meter tape to the siren pole at the ground so that the beginning of the tape is held directly beneath the siren pole and doesn't get pulled out of place by people stepping on it.

Theoretical Background

Physics texts vary quite a bit in their presentation of the theory of sound and sound measurement. Some texts do not address all of the numerous similar terms involved, including sound intensity, sound intensity level, sound pressure and sound pressure level. Additionally, the theory of sound measurement may be less familiar to some physics teachers than more common topics such as mechanical energy or electromagnetism. Accordingly, I am going to present a lengthier description of the theory here than I have in the other chapters.

Our first point of clarification will be to distinguish between *sound intensity level* (β) and *sound pressure level* (SPL, or L_p). Both of these terms refer to logarithms of ratios, and both are expressed in decibels (dB). Physics texts tend to focus on the sound intensity and sound intensity level. Texts focusing on measurement will also address sound pressure and sound pressure level, and technicians talking about sound measurements will almost always speak of sound pressure level. I will address each of these terms.

Sound pressure is the root-mean-square of the instantaneous air pressure reading, in a specific frequency band over a specific time interval. Sound pressure is measured in pascals, Pa. The sound pressure level in dB is based on a logarithm of the ratio of sound pressure to a reference pressure of 20 μPa, which is the approximate sound pressure at the threshold of human hearing. We will not calculate sound pressure level directly in this experiment, but just for reference the defining equation for sound pressure level is

$$L_p = 10\log\left(\frac{p}{p_0}\right)^2 = 20\log\frac{p}{p_0}$$

where p is the sound pressure in pascals (Pa) and p_0 is the reference pressure of 20 μPa. A sound level meter is a pressure-based instrument, and sound pressure level is what its display indicates. So the correct way to speak about what the sound level meter indicates is to call it the sound pressure level (L_p).

The numerical value of the L_p is equivalent to the numerical value of the sound intensity level, β. The sound intensity level is based on a logarithm of the ratio of sound intensity, I (W/m^2), at a particular location to a reference intensity, I_0, equal to 10^{-12} W/m^2, which is approximately the sound intensity at the threshold of human hearing. The sound intensity level, in decibels (dB), is calculated as

$$\beta = 10\log\frac{I}{I_0}$$

The sound intensity level, and the equation above, is what the students will use in their analysis for this experiment. The important point here is that the readings from the sound level meter are sound pressure level values, but they are numerically equivalent to the values we obtain from the sound intensity level equation. So we can develop our theory based on the relatively simple concept of sound intensity, but we can collect our data from a sound level meter that measures the sound pressure level.

Now let's back up a bit and establish the idea of sound intensity. We consider an ideal, isotropic sound source, that is, a source that radiates acoustic energy equally in all directions, as represented by the diagram below. Sound energy propagates from the source in a spherical distribution. If the source radiates sound energy with a total power of P watts, then

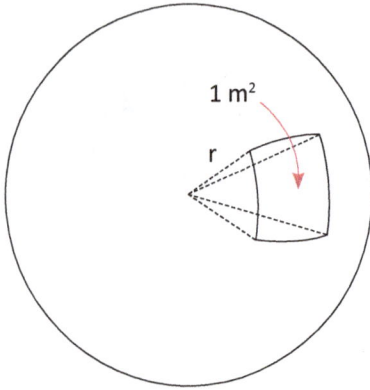

1 m²

r

At a distance *r* from an isotropic sound source, the amount of power passing through a 1- square meter area on the surface of the wave front is $P/(4\pi r^2)$.

at a distance r_1 from the source this total power will be distributed evenly around the surface of a sphere centered at the source. The surface area of a sphere is $4\pi r^2$, so that the surface power density in watts per square meter at the sphere, which is called the intensity, *I*, will be

$$I_1 = \frac{P}{4\pi r_1^2}$$

Similarly, at a larger radius, r_2, the intensity will be

$$I_2 = \frac{P}{4\pi r_2^2}$$

The ratio of these two intensities will be

$$\frac{I_2}{I_1} = \frac{\dfrac{P}{4\pi r_2^2}}{\dfrac{P}{4\pi r_1^2}} = \frac{r_1^2}{r_2^2}$$

Solving this equation for I_2 gives

$$I_2 = I_1 \left(\frac{r_1^2}{r_2^2} \right)$$

This last equation displays the inverse-square behavior of intensity vs. distance. The equation allows us to calculate the intensity at a location r_2 from a known reference intensity at location r_1.

Of course, applying this theoretical development successfully to this experimental scenario depends on a couple of simplifying assumptions. The first is the assumption that we are using an isotropic sound source. No sound source is perfectly isotropic, and some sound sources, such as those with large horns, are very directional. But as it turns out, the design of a piezo siren is such that the siren can be modeled with reasonable accuracy as an isotropic sound source.

A second assumption is that sound energy arriving at the measurement location is coming directly from the source, and not from reflections. Out in a field far from any buildings, the only sources of any significant reflected sound energy are the ground and the people performing the experiment. As it turns out, grassy fields and humans wearing soft clothing are both pretty good absorbers of sound energy in the frequency range produced by the siren. This means that as long as you stay away from hard surfaces such as concrete, hard packed dirt, and buildings, and as long as the students are wearing soft clothing (as opposed to, say, space suits), it is reasonable to assume that the amount of reflected sound energy arriving at the sound level meter is minimal.

Analysis

In this experiment the goal is to use the SPL measurement at a distance of one meter as a reference (r_1) for predicting the intensities at other distances. To do this students will use the mean SPL measurement at the one-meter location as the value of β at one meter. Then they will solve the sound intensity level equation for I to get the sound intensity (W/m^2) at the one-meter location, I_1. With the value of I_1 and the distances at the other measurement locations, students can use the inverse square law above to calculate the intensities in W/m^2 predicted for the other 10 measurement locations. Next, students will use the defining equation for β to convert each of the predicted intensities into a sound intensity level in dB.

At this point the students have 11 predicted SPL values in dB and 11 means of the measured SPL values at various distances, also in dB. (The predicted and measured values at 1 meter are identical since the measurement there was used as β for that location.) They will calculate the prediction-result difference ratio for each location. They will also construct a plot of SPL vs. distance showing both predicted and experimental values (the means of measurements at each distance). Students should compute the sample standard deviation (s) of the SPL values at each location to use as an estimate of the uncertainty in the SPL at each location, and use these uncertainty figures to generate error bars on the experimental curve. If the experiment is successful, the experimental curve will closely track the predicted curve, and the prediction at each distance will fall within the range of experimental uncertainty as indicated by the error bars.

It is no secret that students often struggle when operating in a study involving logarithms. There is an important aspect of this experiment that can help shed light on logarithmic variation, so I always like to bring it up and discuss it with the students.

Human hearing is effective over 12 orders of magnitude of sound intensity, an enormous range. Not only so, but the nature of our hearing is logarithmic, which means, for example, that we perceive a doubling of the intensity from 10^{-12} W/m^2 to 2×10^{-12} W/m^2 as representing the same perceived change in loudness as a doubling from 10^{-4} W/m^2 to 2×10^{-4} W/m^2. In the first instance the increase in intensity is 0.000000000001 W/m^2, and in the second instance the increase is 0.0001 W/m^2, and yet we perceive these two increases to represent the same degree of change in loudness.

From the ideas developed in the Theoretical Background section, it is easy to show that a doubling of intensity represents an increase in intensity level of 3.01 dB (that is, 10 log2 = 3.01). Now let's consider what happens if wind or ambient noise were to cause fluctuations in intensity of a factor of two at a location where the measured SPL is 80 dB. An SPL of 80 dB converts to an intensity of 0.0001 W/m^2. A fluctuation of a factor of two in the intensity that gave us a measurement converting to 0.0002 W/m^2 when our prediction was 0.0001 W/m^2 would mean that in terms of intensities the measurement error would be 100%. But when considered as intensity levels, the same fluctuation would be 83 dB compared to a prediction of 80 dB, giving a measurement error of only 3.75%!

Again, our hearing perception works logarithmically. Expressing the difference between prediction and measurement in terms of intensity levels in dB captures the way our perceptions work a lot better than the difference between intensities does. For this reason, students should perform their error analysis with intensity levels/sound pressure levels in dB. If in your class discussions you can lead students to understand these last three paragraphs, it should pay dividends for them down the road as they asymptotically approach comprehension of what a logarithm is.

Time Averaging and Weighting Curves

The sound level meter has a switch for FAST/SLOW time averaging of the meter reading. This switch should be set to SLOW to minimize the fluctuations in the readings due to wind. There will also be a switch to activate A-weighting or C-weighting. Set this switch to the A-weighting position, which is typical for noise measurements in industry. With advanced or honors students I enjoy enriching their experience in the area of sound measurement by explaining this function, so I will explain it here in the next couple of paragraphs. For non-honors students the explanation will probably qualify as too much information. It is not really a critical part of the experiment, so you if decide your students should just stick to the task at hand, then set the weighting switch to A-weighting and forget about it.

Human hearing is most sensitive in the center of the frequency range of human hearing, from around 300 Hz to 8 kHz (and especially sensitive from 1 kHz to 6 kHz). Outside of this range our hearing sensitivity falls off rapidly. For example, for a human to hear a sound at 40 Hz, the intensity (which is the power density at the surface of the expanding wave front in air) of the sound must be about 10,000 times higher than it must be for a sound at 1,000 Hz. This is because the threshold of human hearing at 40 Hz and 1,000 Hz is at intensities of about 10^{-7} W/m^2 and 10^{-11} W/m^2, respectively.[2] This is a difference of four orders of magnitude, or 40 dB. When A-weighting is selected on the sound level meter, the meter will roughly mimic the frequency response of human hearing, which means the meter will be much less sensitive to sounds below 300 Hz or above 8 kHz. (These frequencies are very approximate. The farther out toward the extremes of the human hearing range the sound is, the less sensitive the meter will be.)

The figure on the next page shows the weighting curve used by the sound level meter when A-weighting is selected. The frequency response of human hearing follows a similar (but far less regular) curve. As you can see from the graph, the difference in the relative response at 40 Hz and 1,000 Hz is about 40 dB. So when A-weighting is selected, the meter's sensitivity mimics that of the human ear. (We won't worry here about what the C-weighting curve is used for.)

So here is how the A-weighting curve helps in the case of our experiment. The siren, which is specifically designed to attract the attention of humans, emits a warble tone at frequencies right in the center of the human hearing range, where our ears are most sensitive. If the sensitivity of the sound level meter is set to match the human hearing response (which is what A-weighting approximates) then the meter will be less sensitive to low-frequency noise that may be present in the environment. So if a truck rumbles by, making a lot of low frequency noise, the meter will not hear it as easily. The bottom line is that the A-weighting curve helps the meter listen to the siren rather than the rumble. Of course, the meter will still be sensitive to all sounds that fall within the boundaries of the weighting curve, such as chirping birds or children screaming at a nearby playground. The weighting curve will not help you there.

Pre-Lab Discussion

Review these items with the group on the day of the experiment, prior to taking data.

---

2   These values are approximate because, obviously, the sensitivity of human hearing varies from person to person.

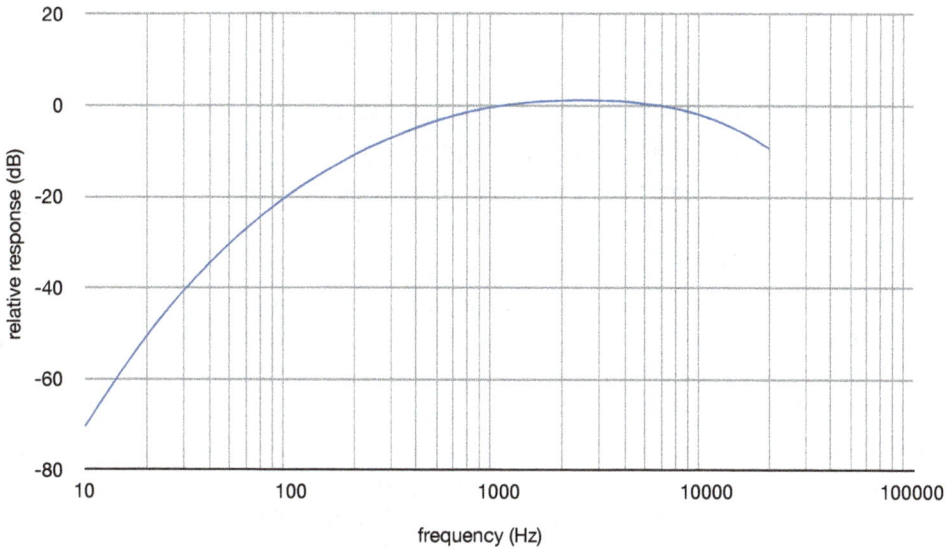

The relative frequency response produced by the A-weighting curve in a sound level meter. The frequency response simulates that of human hearing.

1. We will be working close to a VERY loud siren, and the sound from it is very unpleasant in the absence of hearing protection. Everyone will be provided with a pair of ear plugs to wear while we are taking data. (The students will be interested in this, primarily because of the novelty of needing to wear ear plugs. Their interest will be much higher if the ear plugs you buy have day-glow colors and are connected together with a plastic cord. See the Grainger part number listed in the materials list.)

2. Students need to be shown the correct way to insert ear plugs, which is usually illustrated on the package. It goes like this for the right ear: Roll the ear plug in the fingers of both hands to squish down the diameter. Hold the ear plug tightly in the fingers of your right hand to keep it from expanding. Extend your left arm behind your head, grasp your right ear from behind at the top of the ear and pull the ear toward the back of your head. While pulling on the ear with the left hand, insert the plug with the right hand. For the ear plug to insert correctly you must angle it toward the front of your head (that is, point it more or less toward your nose during insertion). This all takes practice and students may need to try it several times for proper insertion. The ear plugs will only provide full protection when fully inserted. When fully inserted the ear plugs will not be protruding from the ears at all.[3]

---

3   The teacher will probably care more about proper ear plug insertion than the students. Students don't seem to be bothered as much and tend to regard the difficulty of inserting the earplugs as a hassle. For most students, even if they don't wear the ear plugs at all the noise from the siren is not likely to cause any harm, unless a student has sensitive hearing. Students with any kind of hearing disability or sensitivity should be required to wear the ear plugs fully inserted.

3. SPL readings will be taken at the following distances from the siren: 1 m, 2 m, 3 m, 4 m, 5 m, 6 m, 8 m, 10 m, 12 m, 16 m and 20 m. The measuring tape will be stretched out on the ground to make it easy to locate each measurement location. At each location we need to take at least 4 or 5 measurements (with different students reading the sound level meter) so we have a distribution of data at each location.

4. We will orient our direction on the field so that the sound level meter is pointing away from the wind (that is, pointing in the same direction the wind is blowing), and away from any unavoidable noise sources in the area (such as the sound of an air conditioner on a nearby building).

5. We will assign one student as the first to take data. This student will move down the tape, calling out a reading at each location while another student records the data. Then we will hand the meter to another student and do it again repeating until we have at least four or five data sets.

6. We will need about six of the students not taking the measurements to stand closely together, shoulder-to-shoulder, with three people on each side of the student with the meter. This will make a human wall to help screen the sound level meter from wind and noise. Students should form themselves into an arc to help screen the sound level meter from noise or wind gusts coming from the side, but should not block the line-of-sight between the meter and the siren.

7. While taking data, everyone one must remain still and absolutely silent. If there is any wind or ambient noise, the sound level meter readings will be fluctuating quite a lot. The student taking the reading should watch the meter closely for several seconds at each location and ascertain the lowest possible reading available at that location. This reading will occur during the best possible lull in wind and ambient noise. After studying the readings for a few seconds to get the lowest value that occurs at that location, call out that lowest value for the data recorder to take down. Note that an indication that the data collection is going well will be that the sound pressure level always goes down as we get farther from the siren.

8. To take a reading, the student holding the sound level meter should hold the instrument waist high, point it directly at the siren, and stand over the measurement tape so the tip of the instrument is held directly above the measurement tape at the desired distance from the siren.

References

The A-weighting curve for sound level meters is specified in ANSI standard S1.4.

Student Instructions

A set of instructions you may reproduce and give to students begins after the following illustrations.

The piezo siren taped to the pole, and connected with the alligator clip leads to the battery, which is sitting on the step stool.

Stretching out the measuring tape, and preparing to take measurements.

Taking measurements. One student holds the sound level meter above the tape at the appropriate distance from the siren while other students crowd closely around to help block wind or ambient noise.

Sound
Sound Level Measurement and Inverse-Square Variation

A well-known instance of an inverse square law in nature is that sound intensity from an isotropic sound source decreases inversely as the square of the distance from the source. With this inverse square law and the equation relating sound intensity, I (W/m^2), to sound intensity level, β (dB), one can use the intensity level at one location to predict intensity levels at other locations. And since the numerical value of the sound pressure level (SPL or L_p) is equivalent to the value of the intensity level, SPL field measurements can be compared to intensity level predictions based on a reference intensity at a certain location. This is what we will do in this experiment, which we will conduct together as a class.

Taking Data

1. In a large, open outdoor area (such as a soccer field far away from buildings), lay out a 20-meter measuring tape next to an isotropic (or approximately so) sound source. Measure and record the SPL at a distance of 1.00 m from the sound source. Then make additional SPL measurements at distances of 2 m, 3 m, 4 m, 5 m, 6 m, 8 m, 10 m, 12 m, 16 m, and 20 m from the source.

2. On the sound level meter, set the averaging to SLOW and the weighting to A-weighting.

3. Using different students for the measurements, take at least four or five SPL measurements at each location so that you have a distribution of data at each location. The mean of the values will be the experimental value of the sound pressure level for that location, and you can use the sample standard deviation of the data at each location as an estimate of the uncertainty in the SPL measurement at that location.

4. Since both sound waves and wind are fluctuations in air pressure, instruments that measure SPL are extremely sensitive to wind. Thus, you must take data on a relatively calm day. To help prevent fluctuations in wind from affecting your measurements, use the students in the class as a human shield to screen off the instrument as much as possible from the wind.

5. Obviously, while taking data everyone present must be silent and ambient noise must be minimal.

6. The sound source is a very loud piezo siren. Wear ear plugs while taking data.

Analysis

To expedite your data analysis, use a spreadsheet to calculate the mean and standard deviation for each of your data sets. In Microsoft Excel this is simple to do. The steps are as follows:

1. Enter your data (the SPL values of for one location) into a column in the spreadsheet.

2. Click on an empty cell in the column below the data. From the tool bar select Insert and Function. In the formula builder under the *fx* tab double click on AVERAGE and hit the enter key. You will then have the mean of your data set in the spreadsheet.

3. Click on another empty cell below the data in the same column. Repeat the previous step, but this time select STDEV. A box will appear in the data column that may include the cell containing the mean, and may or may not include your data. Click on the upper and lower right-hand corners of the box and drag the edges so the box includes only the data and not the cell containing the mean. Then hit enter. The sample standard deviation of your data will now appear in the cell.

Use the mean of the SPL data at 1.00 m as a reference to predict intensities at other distances. To do this, convert the mean of the 1.00 m SPL measurements into an intensity (W/m^2). Then use this intensity and the inverse square law to predict the intensities at each of the other distances. Convert each of these predicted intensities into β values, which are equivalent to SPL values. These calculated SPL values will be your predicted values.

Using the predicted and experimental SPL values, calculate the prediction-result difference ratio for each location 2.00 m or more from the sound source. Use a spreadsheet or other application to plot curves showing SPL vs. distance for both the predicted and experimental SPL values. Plot the two curves on the same set of axes. Show error bars (based on the standard deviations) on your experimental curve.[1] Compare the experimental and predicted curves quantitatively (difference ratio at each location, etc.) and qualitatively (shape of the curve) in your discussion. In the event that your experimental curve stays consistently above or below your predicted curve you should try to explain why. The location of the predicted curve relative to the error bars on the experimental curve will be a major factor to consider in your analysis.

There is a reason why we are using values of the sound intensity level β (equal to SPL values) for the error analysis instead of values of the intensity, I, and you should seek to understand what it is. Your instructor will discuss this with you in class. Understanding this issue will really help you appreciate the nature of measurements relating to variables that vary logarithmically.

Terms and Notation

In your report be careful to refer correctly to the technical terms involved in this experiment. As you know from your studies, β is the sound *intensity* level in dB. This parameter is distinct from the sound *pressure* level (SPL or L_p), also in dB, although their numerical values are the same. The distinction is subtle, but boils down to this:

---

1 See *The Student Lab Report Handbook* for details on how to do this in Microsoft Excel. For assistance with other applications, see the free resources at novarescienceandmath.com.

When referring to your calculations of β, refer to the sound intensity level. When referring to measurements from the sound level meter, refer to sound pressure level. In this experiment we are calculating intensities and intensity levels, and we are measuring pressure levels.

Your instructor may or may not opt to discuss with the class the technical background behind our decision to set the sound level meter to A-weighting while taking data. But since your data are A-weighted data, you should denote this with the correct notation, which is to place the designation "(A)" after the dB in the units. Thus, an A-weighted SPL measurement of 92.5 dB should be written as "92.5 dB (A)."

Photogate Notes

As mentioned above, if funding is an issue you can do this experiment with only a stop watch for a timer. But when funds are available you will want to acquire a digital timing system for more accurate timing. The photos beginning on page 49 show two different photogate models in use with the same timing system. Both are made by Daedalon Corp. The EA-27 style photogate has metal ears designed to clamp on to a Daedalon air track. In one of the photos (bottom of page 50) these photogates are taped upside down to the table top for use with the Hot Wheels track. The other photogate shown in most of the photos is the ET-45. Both of these photogate models work just fine, although as shown in the photos the ET-45 photogates require clamps and other hardware for support.

The timer shown in the photos is the Daedalon Corp. model ET-41. The Daedalon equipment is available from Flinn Scientific (flinnsci.com) and from The Science Source (thesciencesource.com). Flinn sells the ET-41 timer with a single EA-27 photogate in a kit, model AP5745. When purchasing from Flinn you have to request adding an additional photogate to the kit. The Science Source lists the ET-41 timer and the EA-27 and ET-45 gates separately. Regardless of the photogate option you use, you will need one timer and two photogates.

Finally, if you use a Daedalon air track elsewhere in your science program (see, e.g., Part 2 Demos 5) you can use the EA-27 photogates with it. This may be a good reason to go with the EA-27 photogates, although I prefer having both types in the lab for use with different experiments.

Materials List

Experiment 1 - The Bull's-Eye Lab

#	Item	Qty	Source	Price
1	steel ball, 1 inch diameter, such as #P1-5003	1	hometrainingtools.com	$2
2	laboratory support rod or ring stand Item# CE-STANDA	1	hometrainingtools.com	$18
3	small clamp such as #DWHT83139	1	Hardware store	$11
4	shelving support rail eg. #FG4A6901WHT	1	Hardware store	$6
5	stop watch	1	general	$5
6	masking tape	1	general	$1
7	plumb bob such as Model # 908	1	Hardware store	$5
8	6' nylon string		general	
9	meter stick	1	general	
10	carpenter's level	1	Hardware store	$10
11	target (photocopies)		see p. 177	
12	carbon paper (one sheet for class)	1	Office supply store	

Experiment 2 - The Friction Challenge (per team)

#	Item	Qty	Source
1	brass plate, 10 in x 4 in x 5/16 in (approx.)	1	industrialmetalsupply.com
2	brass flat bar, 1 in x 1 in x 1.25 in (approx.)	1	industrialmetalsupply.com
3	waterproof polishing paper in four grades: 120-C, 220-A, 320-A, and 400-A, such as part nos. 19823, 19808, 19798, and 19795	1 ea	abrasivesales.com
4	cleaning cloths		general store
5	nylon cord		general store

71

Materials List

6	WD-40 spray silicon lubricant		general store
7	triple-beam balance, such #02-7600	1	arborsci.com	$89
8	other standard lab equipment such as pulleys, table clamps, mass sets, adjustable ramp with angle gauge, timing equipment and measurement tools		read the experiment procedure to decide which of these you want to use.

Experiment 3 – Rotational Kinetic Energy

1	steel ball, 1 inch diameter, such as # P1-5003	1	hometrainingtools.com	$2
2	shelving support rail eg. #FG4A6901WHT	1	Hardware store	$6
3	stop watch (only if digital timing system is not available, item #14)	1	general store	$5
4	masking tape		general store	
5	carpenter's level	1	hardware store	
6	track support frame (described in experiment, made of aluminum stock)	1		
7	J.B. Weld epoxy adhesive		hardware store	$4
9	meter stick	1	general store	
10	drafter's triangle	1	hardware store	
11	copy paper (for leveling)			
12	micrometer	1	hardware store	$20
13	dial caliper, Model No. T9F534164	1	hardware store	$25

optional equipment if digital timer system is used:

Materials List

14	Combination digital timer and photogates such as #P4-1450	1	arborsci.com	$279
13	support stand, such as #66-4220	2	arborsci.com	$18
14	clamp holder, such as #66-8290	3	arborsci.com	$17
Experiment 4 - Calorimetry				
1	aneroid dry calorimeter, such as #CE-CALORIM	2	arborsci.com	$12
2	specific heat cylinders, such as this set	2 or 1 set	onlinesciencemall.com	$20
3	thermometer	3	general store	$3
4	needle-nose pliers	1	hardware store	$15
5	laboratory oven (or household oven)			
6	freezer			
7	fishing line		general store	$1
8	triple-beam balance, such #02-7600	1	arborsci.com	$89
9	J.B. Weld epoxy adhesive		general store	$4
Experiment 5 - Sound				
1	sound level meter, such as #P7-7700	1	arborsci.com	$99
2	piezo siren, such as Radio Shack Model 273-079	1	radioshack.com	$7
3	measuring tape		hardware store	$5
4	lantern battery, 6V	1	hardware store	$11
5	duct tape		hardware store	

Materials List

6	copper tubing, 1/2 inch x 6 feet long. Any other lightweight pole or pipe will work just as well.	hardware store	varies
7	alligator clip test lead, 36 in, such as Allied Stock No. 70209822 (this stock number includes a pair of leads, one red and one black) such as SKU143292	2	banggood.com	$2
8	folding step stool	1		
9	disposable ear plugs, available at drug stores, or industrial style such as Grainger Item No. 6XF60	1 set per person	grainger.com	$4

This Appendix is a brief primer on measurement and error. To the novice, making a measurement seems like a straightforward task. But in fact, measurement constitutes an entire field of study in itself, the field of metrology. Making measurements is a complex undertaking.

Experimental science deals in measurements. Most of the raw data associated with or resulting from an experiment are measurements of some kind. For this reason, it is essential that all high school science students know how to make proper measurements, and know the limitations of measurements with respect to accuracy and precision.

For the present work our goal is simply to present the basic issues surrounding measurement and error that high school students will need to know. Much of the material in this Appendix is adapted from my introductory physical science text, *Accelerated Studies in Physics and Chemistry*, and my handbook for all high school science students, *The Student Lab Report Handbook*.[1]

A Note About Experimental Error

One of the conventional calculations in high school science labs is the so-called "experimental error." This experimental error is typically defined as the difference between the predicted value and the experimental value, expressed as a percentage of the predicted value, or

$$\text{experimental error} = \frac{|\text{predicted value} - \text{experimental value}|}{\text{predicted value}} \times 100\%$$

From the perspective of the average high school student, this use of "experimental error" makes perfect sense. After all, student are studying well-established theories and the goal of the experiment is to learn about the theory, not to validate or refute it. In the world of science, however, experiments are the golden standard by which theories are judged. When there is a mismatch between theory and experiment, it is often the theory that is found wanting. That is how science advances.

In my early books, such as *The Student Lab Report Handbook*, I used this same terminology ("experimental error") to express the difference between prediction and result. Over the years, however, research and discussions with practicing scientists have led me to the conclusion that this terminology is misguided. Used in this way the term *error* implies that the theory is *correct* and that the error in the experiment may be summarized by this difference equation. However, the difference between prediction and experimental result may not be caused by deficiencies in the experiment. In more general scientific practice the theory may *not* be correct. Thus, in secondary classrooms it is better to reserve the term *error* for discussions about lack of accuracy in specific measurements, when the measurement is known to contain or is suspected of containing error (that is, differing from the true value,

---

1   Visit novarescienceandmath.com for details.

see below). Referring to the overall difference between prediction and experimental result as "experimental error" is a bad habit to get into.

Consider this case: an experimental measurement of velocity produces a value that is consistently less than the predicted value. Most likely this is because the predictions did not take air resistance into account. Is this an experimental error? It is more correct to say that the theory is inaccurate because we made the unrealistic assumption that there would be no air resistance. Such causes of differences between predictions and measurements are quite common, and it is great if the future scientists in your class can understand that this is not an error in the experiment.

As a result of these considerations, I have adopted a different convention. I now use the phrase "percent difference" to describe the value computed by the above equation. When quantitative results can be compared to quantitative predictions, students should compute the percent difference as

$$\text{percent difference} = \frac{|\text{predicted value} - \text{experimental value}|}{\text{predicted value}} \times 100\%$$

In the Discussion section of their lab reports, students should state the value(s) of the prediction-result difference ratio for their experiment. After doing so, much of their subsequent analysis of the experimental result will consist in attempting to identify the reasons for this difference. Students may use the possibility of *errors* in different measurements, along with other factors such as lurking variables or insufficiently elegant experimental methods, in their attempts to account for the prediction-result difference. Being able to explain the prediction-result difference for an experiment is one of the most important jobs of the scientist—and the science student.

Types of Error

There is no such thing as a perfect or exact measurement. All measurements entail limitations that cause the measured value of a physical quantity to differ from the true value of that quantity. The limitations in a measurement can derive from inaccuracy (error), imprecision, or both.

The error present in scientific measurements can be divided into two main categories, random error and systematic error. Random errors are caused by unknown and unpredictable fluctuations in the experimental setup. These fluctuations influence measurements in a random fashion. Here are some examples of factors contributing to random error:

- Changes in the apparatus or instrumentation due to temperature fluctuations that cause materials to expand and contract.

- Vibrations or air currents that influence the measurements.

- Slight fluctuations in a measured value due to minute variations in equipment or instrumentation alignment.

- The presence of dust or contaminants that influence a measurement.

- Electronic noise that influences the readings in electronic instruments.

When a scientist calculates and analyzes the uncertainty in a measurement, it is the random error that is under consideration. Random error causes a series of measurements to fluctuate randomly around the mean value of the measurements. Whether or not the random error is noticeable or detectable depends on the precision or resolution of the instrument used to make the measurement. (We will discuss precision in more detail shortly.) If numerous measurements of a particular variable are performed with an instrument sensitive enough to detect the fluctuations in the measurement due to random error, the measured values will generally form a Gaussian distribution about the mean of the measurements, as depicted in the diagram to the right. While not uni-versally accepted, it is common practice to use the sample standard deviation of the measurements as an estimate of the *uncertainty* in the measurements.

Systematic errors are errors that bias the experimental results in one direction. Systematic error can be caused by equipment defects, miscali-bration of measurement instruments, or an experimenter who consistently misreads or misuses the instruments in the same way. Usually when discussing systematic error we are talking about problems that could be eliminated by proper use, calibration and operation of the equipment.

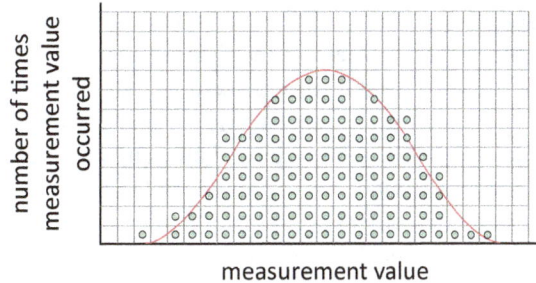

° = one occurrence of a particular measurement value

A Gaussian distribution forms from the data when a specific measurement is made repeatedly with an instrument precise enough to show the variation.

Systematic error can also occur if there is a lurking variable affecting all of the measurements, such as gravity or magnetism. Such factors can bias all of an experimenter's measurements in the same way. Yet another way systematic error can creep into an experiment is by factors that are left out of the theoretical predictions, and that result in a prediction-result difference ratio biased in a certain direction. A common example of this for high school physics is when experiments in mechanics do not take friction into account. This is another type of systematic error, although this time it is not an error with operating the equipment, it is an error built-in to all of the experimenter's predicted values.

Accuracy and Precision

The terms accuracy and precision refer to the limitations inherent in making measure-ments. Every measurement contains error, which is a limitation on the accuracy of the measurement. Every measurement also has a limitation on its precision.

Accuracy relates to error, which is the difference between a measured value and the true value. The lower the error is in a measurement, the better the accuracy. As mentioned above, errors can be random or systematic, and there are many sources of error, including human mistakes, malfunctioning equipment, incorrectly calibrated instruments, uncontrollable random fluctuations, or unknown factors that are influencing a measurement without the knowledge of the experimenter.

Precision refers to the resolution or degree of "fine-ness" in a measurement. The limit to the precision that can be obtained in a measurement is ultimately dependent on the instrument being used to make the measurement. If an experimenter wants greater precision, he cannot achieve it simply by being more careful. Instead, he must use a more precise instrument. The precision of a measurement is indicated by the number of significant digits (or significant figures) included when the measurement is written down (see below).

Here is an example to illustrate the difference between accuracy and precision. Let's say you have a bathroom weigh scale that is very accurate. If you measure your weight on it you can be assured that the weight reading will agree with a measurement made with the finest and most carefully calibrated instrument. But what if the bathroom scale only reads your weight to the nearest pound? This rounding to the nearest pound is a limit on the precision of the scale. So this scale is accurate, but not very precise.

Now imagine that you have a friend that has a different type of scale that reads weights to the nearest 1/100th of a pound but is manufactured according to a poor mechanical design. This scale is much more precise, and can give 5-digit weight readings instead of the 3-digit readings the other scale gives. However, the more precise scale might be quite inaccurate. The poor mechanical design might mean that the scale gives you 5-digit readings that are always a few pounds too high or too low. This scale is significantly more precise, but less accurate.

This example shows how accuracy and precision are interrelated. The potential of an accurate mechanism or measurement technique is wasted if the precision of an instrument is insufficient, and high precision is wasted if the equipment and measurement techniques introduce so much error as to render the precision meaningless. In fact, as a practical rule of thumb, we want the precision of an instrument to be sufficient to make manifest the variation in the measurement due to random error.

One more example will further illustrate the distinction between accuracy and precision. Consider measuring the diameter of a quarter using a common child's 12-inch rule marked off in tenths of an inch. The ruler gives us the tenths with certainty, and we have to estimate the hundredths. The quarter appears to be just a little bit past the halfway mark between 0.9 inches and 1.0 inches. Estimating as carefully as possible the measurement looks like 0.96 inches, a value with two significant digits. This measurement is as accurate as it can be, meaning that the measurement is limited by its imprecision, and not by error in the measurement process. The U.S. Mint website specifies the diameter of a quarter to be 0.955 inches, or 24.26 mm. These are more precise values, with three and four significant digits, respectively.

If we use the conversion to inches of 25.4 mm = 1.0000 inches (which is the contemporary definition of an inch, and is thus exact), the 24.26 mm converts to 0.9551 inches. But since all of the dimensions in inches on the U.S. Mint website stop at thousandths, I assume that this is the limit on the manufacturing tolerances used for making quarters.

Parallax Error and Liquid Meniscus

We will address correct measurement practices in detail in a later section, but to facilitate discussion of the examples in the next section we need to address two common issues here, avoiding *parallax error* and working correctly with the *meniscus* on liquids. These terms both have to do with using analog instruments with measurement scales that must be correctly aligned for an accurate measurement.

Parallax error occurs when the line of sight of a person taking a measurement is at an incorrect angle relative to the instrument scale and the object being measured. As shown

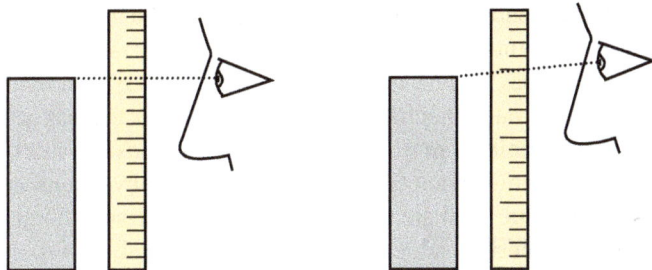

In the sketch on the left the measured object, measurement instrument and viewer line of sight are correctly aligned. In the sketch on the right the misalignment of the viewer is causing parallax error in the measurement.

in the figure, the viewer's line of sight must be parallel to the lines on the scale, and perpendicular to the scale itself. Misalignment of the viewer's line of sight will result in a faulty measurement due to parallax error.

When a liquid is placed in a container the surface tension of the liquid causes the liquid to curve up or down at the walls of the container. In most liquids the surface of the liquid curves up at the container wall, but a well-known example of downward curvature occurs when liquid mercury is placed in a glass container. In this case the surface of the liquid metal curves down at the container wall. We will concentrate here on the common upward curvature exhibited by water.

If the container is tall and narrow, as with a graduated cylinder, then the curving liquid at the edges gives the liquid surface an overall bowl shape. This bowl shaped surface is called a meniscus. The correct way to read a volume of liquid is to read the liquid at the bottom of the meniscus. The next sketch illustrates the correct way to read a volume of liquid in a graduated cylinder by reading the liquid level at the bottom of the meniscus and avoiding parallax error.

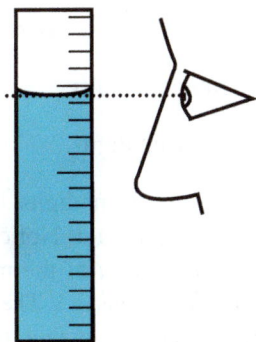

Correctly reading a liquid volume in a graduated cylinder. The viewer's line of sight is at the bottom of the meniscus, and is perpendicular to the scale to avoid parallax error.

Significant Digits

The notion of the precision of a numerical value enters into the physical sciences in several instances. First, we routinely deal with measurements, and every measurement has an inherent precision that depends on the instrument used to make the measurement. Second, we often use physical constants such as the acceleration due to gravity or the speed of light. Seldom do we use *exact* values for such quantities (if such can even been said to exist); generally we use an approximate value that has adequate precision for the task at hand. Third, we must constantly deal with units of measure and with conversions of values from one set of units to another. Often the conversion factors we use for unit conversions are not exact, and again we tend to rely on values that are precise enough for the task at hand.

The precision in a numerical value is indicated by the number of so-called *significant digits* (also called significant figures) it contains. Thus, the number of digits we write in any value we deal with in science is very important. In the case of a measurement, the number of digits in the value of the physical quantity is meaningful, because it shows the precision that was present in the instrument used to make the measurement. The same thing applies to physical constants (most of which are themselves measurements) and conversion factors. To express these numerical values correctly, and to use them correctly in subsequent calculations, we must correctly manage the significant digits in three distinct cases:

- interpreting a value someone has already recorded
- reading and recording a new value from a measurement instrument, and
- using values from measurements in computations.

Let's begin by reviewing the rules for determining the precision in a value already recorded, such as a value stated in a text or experimental result. Since the precision is indicated by the number of significant digits, identifying the precision inherent in the numerical value is equivalent to determining how many significant digits there are in the value. The rule for determining how many significant digits there are in a given value is as follows:

> *The number of significant digits (or figures) in a number is found by counting all the digits from left to right beginning with the first nonzero digit on the left. When no decimal is present, trailing zeros are not considered significant.*[2]

Let's apply this rule to the following values to see how it works.

15,679   This value has 5 significant digits.

21.0005   This value has 6 significant digits.

37,000   This value has only 2 significant digits, because when there is no decimal trailing zeros are not significant. Notice that the word *significant* here is a reference to the *precision* of the measurement, which in this case is rounded to the nearest thousand. The zeros in this value are certainly *important*, but they are not *significant* in the context of precision.

0.0105   This value has 3 significant digits, because we start counting with the first nonzero digit on the left.

0.001350   This value has 4 significant digits. Trailing zeros count when there is a decimal.

---

2   This definition is quoted from *Trigonometry*, Charles McKeague and Mark Turner, 6th ed.

The significant digit rule enables us to distinguish between two measurements like 13.05 m and 13.0500 m. Mathematically, of course, these values are equivalent. But they are different in what they tell us about the process of how the measurements were made. The first measurement has 4 significant digits. The second measurement is more precise. It has 6 significant digits, and was made with a more precise instrument.

Consider again the zeros at the end of 37,000 that were not significant. Here is one more way to think about significant digits that is helpful when teaching students about significant digits. The precision in a measurement depends on the instrument used to make the measurement. If we express the measurement in different units, this has no effect on the precision in the measurement. A measurement of 37,000 grams is equivalent to 37 kilograms. Whether we express this value in grams or kilograms, the value still has 2 significant digits.

Now that we have defined significant digits and have seen how to identify them in a numerical value, we will next address how to apply the idea of significant digits when making new measurements.

The issue of which digits are significant with a digital instrument can get very tricky, and depends in a complex way on the design of the sensors, the electronics, and the software or firmware used in the instrument to generate the display. But for the digital instruments commonly found in high school science labs a basic working rule of thumb is to assume all of the digits are significant except the leading zeros.

For analog instruments, here is the rule for determining how many significant digits there are in a measurement one is making:

The significant digits in a measurement are all of the digits known with certainty, plus one digit at the end that must be estimated between the finest marks on the scale of the instrument.

The photos on the next page illustrate this point with two different types of measurements. The photograph on the left side of this figure shows a rule being used to measure the length of a brass block in millimeters (mm). We know the first two digits of the length with certainty; the block is clearly between 31 mm and 32 mm long. We have to estimate the third significant digit. The marks on the rule are in 0.5 mm increments. Comparing the edge of the block with these marks I would estimate the next digit to be a 6, giving a measurement of 31.6 mm. Two digits of this measurement are known with certainty, the third one was estimated, and the measurement has three significant digits.

The photograph on the right side of this figure shows a liquid volume measurement in milliliters (mL) being made with a graduated cylinder. You can see the curvature of the meniscus, which appears reddish in color. Reading the level at the bottom of the meniscus, we know the first two digits of the volume measurement with certainty, because the volume is clearly between 82 mL and 83 mL. We have to estimate the third digit, and I would estimate the line to be at 40% of the distance between 82 and 83, giving a reading of 82.4 mL.

Finally, we will consider the rules for using significant digits in computations, which include any unit conversions that must be performed. First I will list the two rules students must attend to for multiplication and division. These two rules are easy to understand, and it is appropriate for most students to learn to use these rules as freshmen.

Rule 1   Count the significant digits in each of the values you will use in a calculation, including the conversion factors you will need to use. (Conversion factors that are exact are not considered.) Determine how many significant digits there are in the least

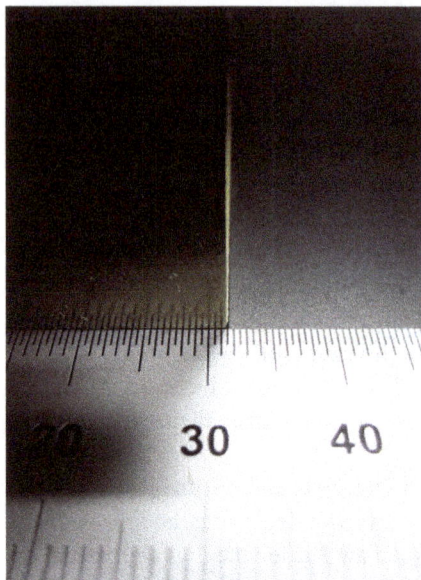

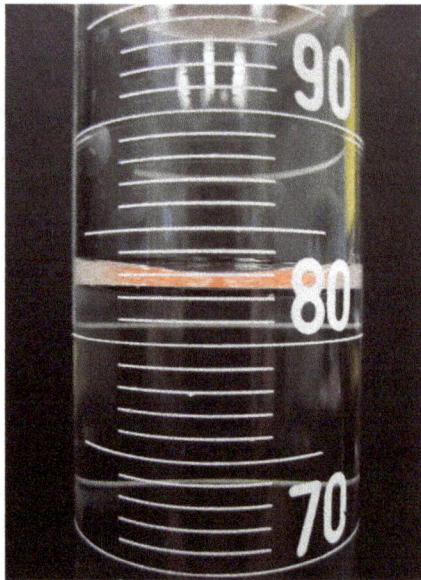

Reading the significant digits on a rule and graduated cylinder.

precise of all of these values. The result of your calculation must have this same number of significant digits.

Rule 2   When performing a multi-step calculation you must keep at least one extra digit during intermediate calculations, and round off to the final number of significant digits you need at the very end. This practice will make sure that small round-off errors don't add up during the calculation. This extra digit rule also applies to unit conversions performed as part of the computation.

For computations involving addition and subtraction a separate rule applies. This rule is a bit more difficult for students to grasp, and I usually recommend that the appropriate place for students to learn to use it is in their first full chemistry course. The rule is cumbersome to put into words, but the example that follows will clarify.

Rule 3   Observe the right-most digit in each value to be added and identify the one that has the highest place value. This place value will be the place value of the right-most digit in the sum.

For example, consider the following sum:

$$
\begin{array}{r}
56.55 \\
2.3626 \\
+14.0 \\
\hline
72.9 \\
\end{array}
$$

The place values of the right-most digits are hundredths, ten-thousandths, and tenths. The highest of these place values is the tenths. Thus, the place value of the sum must be tenths, giving the result shown. The justification for this rule is that the third value (14.0) is only known to the nearest tenth, so it is not possible to state the sum more precisely than to the nearest tenth.

One final note about significant digits I have found helpful is this helpful fact: When a measurement is written in scientific notation the digits that are written down in front of the power of 10 (the stem, also called the mantissa) *are* the significant digits. Sometimes, the only way to write a value with the correct precision is to write it in scientific notation. For example, given a value such as 100 m/s that was precise to the nearest one m/s, the value would need to be written to indicate that it had three significant digits. The only way to do this while using the same units of measure is to write 1.00×10^2 m/s.

Proper Measurement Procedures

There are, of course, many procedures associated with taking different types of measurements with different types of instruments. Earlier in this Appendix we reviewed parallax error and how to avoid it, and the correct way to read a liquid volume at the bottom of the meniscus in a graduated cylinder. And, of course, we just completed a section on correctly reading the significant digits when making measurements. In this section we will limit our discussion to a few additional techniques associated with common measurements students make using standard apparatus in a typical school science program.

Measurements with a Meter Stick or Rule

1. For maximum accuracy, avoid using the end of a wooden rule. The end is usually subject to a lot of pounding and abrasion, which can wear off or compress the wood on the end.

2. As indicated in the accompanying photos, arrange the rule against the object to be measured so that the marks on the scale come in contact with the object being measured. This will help minimize parallax error.

incorrect   correct

Proper placement of a rule or meter stick.

3. As indicated in the figure on the next page, use a straightedge to assure the end of a metal rule is accurately aligned with the edge of an object being measured.

Use of a straightedge for proper alignment of the end of the rule with the object being measured.

Measurements with a Triple-Beam Balance

1. Calibrate the balance before making measurements. This is accomplished by turning the calibration weight under the pan until the scale's alignment marks are perfectly aligned.

2. Make sure the 10-g and 100-g weights are locked into a notch on the beam. Otherwise the measurement will not be correct.

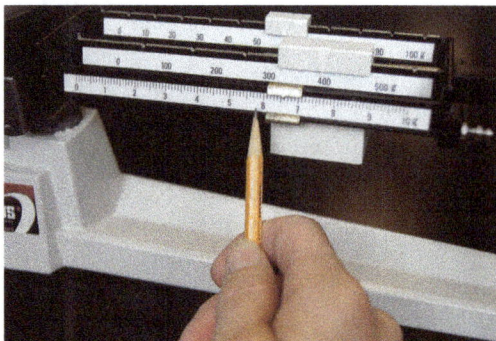

3. When adjusting the position of the gram weight, it is good practice to slide this weight with the tip of a pencil held below the beam instead of with one's finger. If done carefully, this technique will allow the gram weight to be manipulated into position without disturbing the balance of the beam as the balance point is approached.

Move the gram slider with a pencil to prevent disturbing the beam as the balance point is approached.

Measurements with a Thermometer

1. Mercury thermometers are more accurate than spirit thermometers. However, if a mercury thermometer breaks you have a real problem cleaning up and disposing of the spilled mercury. Thus, for student use I recommend using only spirit thermometers.

2. When measuring temperatures be sure to notice that the thermometers have a mark indicating the proper degree of immersion for the most accurate reading.

3. Thermometer accuracy can be severely compromised if gaps get into the red liquid. Always store spirit thermometers vertically in an appropriate rack to help prevent gaps from getting into the liquid.

Units of Measure

SI and USCS Unit Systems

There are many systems of units, but two of them are standard systems all students must study. These are the SI (from the French *Système international d'unités*), typically known in the United States as the metric system, and the USCS (U.S. Customary System).

The USCS, used commonly throughout the United States, is very cumbersome. One problem is that there are many different units of measure for every kind of physical quantity. Just for measuring distance, for example, we have the inch, foot, yard, and mile. The USCS is also full of random numbers like 3, 12 and 5,280, and there is no inherent connection between units for different types of quantities, such as the foot and the gallon. One of the most aggravating features of the system is the use of the pound as both a unit of force and a unit of mass. Designating the units of force and mass values as "lbf" and "lbm" does little to remove the confusion in the minds of students (and adults, too, for that matter).

By contrast, the SI system is simple and has many advantages. There is usually only one basic unit for each kind of quantity, such as the meter for measuring length. The exception is the use of both the liter and the cubic meter for volumes. (However, the while the liter is commonly used as a metric unit, it is technically not part of the SI system.)

Instead of having many different unrelated units of measure for measuring quantities of different sizes, prefixes based on powers of ten are used on all of the units to accommodate different sizes of measurements. And units for different types of quantities relate to one another in some way. Unlike the gallon and the foot, which have nothing to do with each other, the cubic meter is 1,000 liters. For all of these reasons the USCS is not used much in scientific work. The SI system is the international standard.

A subset of the SI system is the so-called MKS system. The MKS system uses the *meter*, the *kilogram*, and the *second* (hence, "MKS") as primary units. Dealing with different systems of units can become very confusing. But the wonderful thing about sticking to the MKS system is that any calculation performed with MKS units will give a result in MKS units. This is why the MKS system is so handy and why science programs in schools (and practicing scientists!) use it almost exclusively.

Fundamental and Derived Units

In the SI system there are seven fundamental units of measure. These are shown in the table below. All other units of measure are derived from combinations of these seven fundamental units.

Physical Quantity	Unit of Measure	Symbol
length	meter	m
mass	kilogram	kg
time	second	s
electric current	ampere	A
temperature	kelvin	K
luminous intensity	candela	cd
amount of substance	mole	mol

SI Unit Prefixes

Unit prefixes are used in the SI system to simplify references to very large and very small quantities. There are twenty standard prefixes listed in readily accessible references, but I recommend that seven of them should be learned by the freshman year in high school, and two more added to the list (tera- and pico-) for older students such as juniors or seniors in physics. These prefixes are listed in the next table.

My justification for the prefixes teachers should require students to learn is based on the ubiquitousness of their use in technology. We live in the era of nanotechnology, when references to prefixes for fractions down to the nano- level are common in technical literature and news. Likewise, improvements in data storage and communications technologies have made the prefix giga- very common, with tera- not far behind. For these reasons basic technical literacy requires that students know these prefixes.

Prefix	Definition	Symbol	Illustration
centi-	10^{-2}	c	There are 10^2 (100) centimeters (cm) in one meter (m).
milli-	10^{-3}	m	There are 10^3 (1,000) millimeters (mm) in one meter (m).
micro-	10^{-6}	μ	There are 10^6 (1,000,000) micrometers (μm) in one meter (m).
nano-	10^{-9}	n	There are 10^9 (1,000,000,000) nanometers (nm) in one meter (m).
pico-	10^{-12}	p	There are 10^{12} (1,000,000,000,000) picometers (nm) in one meter (m).
kilo-	10^3	k	There are 10^3 (1,000) meters (m) in one kilometer (km).
mega-	10^6	M	There are 10^6 (1,000,000) meters (m) in one megameter (Mm).
giga-	10^9	G	There are 10^9 (1,000,000,000) meters (m) in one gigameter (Gm).
tera-	10^{12}	T	There are 10^{12} (1,000,000,000,000) meters (m) in one terameter (Tm).

SI Derived Units

Within the SI system there are many different units of measure derived from combinations of the seven fundamental units. I don't propose to try to treat all of these derived units here, but in the table on the next page I have listed the derived units that are likely to come up in a high school physical science or physics class.

Unit Symbol	Unit Name	Physical Quantity	MKS Derivation
N	newton	force	$kg \cdot m/s^2$
J	joule	energy	$kg \cdot m^2/s^2$
W	watt	power	$kg \cdot m^2/s^3$
Pa	pascal	pressure	$kg/(m \cdot s^2)$
Hz	hertz	frequency	s^{-1}
V	volt	voltage	$k \cdot m^2/(A \cdot s^3)$
Ω	ohm	electric resistance	$k \cdot m^2/(A^2 \cdot s^3)$
C	coulomb	electric charge	$A \cdot s$
F	farad	capacitance	$A^2 \cdot s^4/(kg \cdot m^2)$
H	henry	inductance	$kg \cdot m^2/(A^2 \cdot s^2)$
T	tesla	magnetic field	$kg/(A \cdot s^2)$
Wb	weber	magnetic flux	$kg \cdot m^2/(A \cdot s^2)$

Interesting Unit Facts

I will end this primer with a few facts about units of measure that I have found to be interesting. I enjoy mentioning these from time to time to my students.

1. In the old days the primary standard for the SI unit of length was a 1-m long platinum bar kept in a vault at the International Bureau for Weights and Measures in Sèvres, France. However, since 1983 the standard has been that one meter is equal to the distance traveled by light in a vacuum in 1/299,792,458 of a second.

2. The reference standard for the kilogram is a one-kilogram platinum mass kept in a vault at the International Bureau for Weights and Measures. Officials are discussing a new way of defining the kilogram that may be adopted in the near future.

3. The definition above for the meter immediately raises the question of how the second is defined. Since 1967 it has been defined in atomic terms, specifically, 9,192,631,770 periods of a certain wavelength of light emitted by cesium atoms. By defining the meter and the second in terms of light and atoms the primary reference standards are available to anyone anywhere who has the technology, and it is no longer necessary to use the length of a metal bar in France as a standard that other lengths have to be compared to.

4. The historical length of 1 inch was so close to being 2.54 cm that back in 1959 the powers that be decided to redefine the inch to be equal to *exactly* 2.54 cm. This gave us an exact and easy-to-remember conversion factor to use for converting units between the SI and USCS systems., 1 in = 2.54 cm.

www.ingramcontent.com/pod-product-compliance
Lightning Source LLC
Chambersburg PA
CBHW081552220326
41598CB00036B/6646